Schriftenreihe des Energie-Forschungszentrums Niedersachsen (EFZN)

Band 69

Das EFZN ist ein gemeinsames wissenschaftliches Zentrum der Universitäten:

Numerical study of underground CO_2 storage and the utilization in depleted gas reservoirs

Dissertation

to be awarded the Degree of

Doctor of Engineering (Dr.-Ing.)

submitted by

M.Sc. Cheng Cao

from Sichuan, P.R. China

approved by the Faculty of

Energy and Economic Sciences,

Clausthal University of Technology

Date of oral examination

February 17, 2021

Bibliographical information held by the German National Library

The German National Library has listed this book in the Deutsche Nationalbibliografie (German national bibliography); detailed bibliographic information is available online at http://dnb.d-nb.de.

1st edition - Göttingen: Cuvillier, 2021

Zugl.: (TU) Clausthal, Univ., Diss., 2021

Dean

Prof. Dr. rer. nat. habil. Bernd Lehmann

Chairperson of the Board of Examiners

Prof. Dr. rer. nat. Hans-Jürgen Gursky

Supervising tutor

Prof. Dr. –Ing. habil. Michael Zhengmeng Hou

Reviewer

Prof. Dr. –Ing. habil. Michael Kühn

Nonnenstieg 8, 37075 Göttingen, Germany

Telephone: +49 (0)551-54724-0

Telefax: +49 (0)551-54724-21

www.cuvillier.de

1st edition, 2021

This publication is printed on acid-free paper.

ISBN 978-3-7369-7386-2

eISBN 978-3-7369-6386-3

Acknowledgements

I would like to thank all the people who kindly offered me help during my PhD study at the Research Center Energy Storage Technologies, Clausthal University of Technology, Germany.

First and foremost, I would like to express my deepest appreciations to my supervisor Prof. Dr. -Ing. habil. Michael Zhengmeng Hou for his consistent support, motivation and guidance. It would be impossible for me to study in Clausthal University of Technology and finish this thesis without his support and helpful supervision. I would also like to thank Prof. Dr. -Ing. habil. Michael Kühn for his evaluation and precious comments on this dissertation.

I would like to thank the China Scholarship Council for scholarship support (CSC No. 201808080067).

I own many thanks to all my colleagues at Research Center Energy Storage Technologies and Clausthal University of Technology for a wonderful time. Especially, I would like to thank Dr. Jianxin Liao for his help in solving the problems related to the numerical simulations.

Finally, I want to express my appreciation to my families who constantly give me their total support through all these years. Especially, I would like to give my thanks to my wife Ying Zhou and my son Qingyue Cao. They inspired me to study very hard and finish my thesis as soon as possible, so I can take care of them by their sides.

Cheng Cao

Goslar, October 2020

Abstract

The emission of atmospheric CO_2 is the main contributor to global warming and climate change. Carbon capture and storage (CCS) is considered as the most promising technology for slowing down the atmospheric CO_2 emissions. Meanwhile, CCS is beneficial for the circulation carbon economy. However, CCS has not been implemented on large scale because of the related risks and the lack of economic incentives. This thesis attempts to focus on these two problems and provide some strategies to address them. Regarding the risks associated with CCS, a parametric uncertainty analysis for CO_2 storage was conducted and the general role of different geomechanical and hydrogeological parameters in response to CO_2 injection was determined. Regarding the financial incentives of CCS operation, this thesis attempts to increase the cost-effectiveness of CCS through co-injecting CO_2 with impurities associated with enhanced gas recovery (CSEGR) and using CO_2 as cushion gas in the underground gas storage reservoir (UGSR).

In order to understand the thermal-hydrological-mechanical (THM) process of CO_2 storage, the THM coupled simulator TOUGH2MP (TMVOC)-FLAC3D was developed. By using the developed TOUGH2MP (TMVOC)-FLAC3D simulator, numerical simulation for hundreds of sampled data was performed for results generated by the Quasi-Monte Carlo method. Based on the simulation results, the general role of different geomechanical and hydrogeological parameters was determined in response to CO_2 injection using distance correlation. In addition, a risk factor was defined to characterize the risks of the caprock due to CO_2 injection. The results showed that the reservoir permeability and the injection rate are the two most important factors in determining the pressure change. Moreover, the reservoir Young's modulus plays the most vital role in formation deformation including vertical displacement. The pressure change exhibits a much closer correlation with the risk factor in comparison to the formation deformation, indicating the importance of pressure change in the integrity assessment of the caprock. By using the machine learning approach in support vector regression (SVR), the SVR surrogate model was well-trained based on the data regarding simulated results, and its reliability was verified using the test data. Thereafter, the formation response including the pressure change as well as formation deformation, can be predicted using the trained SVR surrogate model within a very short time. The methods and working scheme applied in this work can be used to guide time and effort spent mitigating the uncertainty in these parameters to acquire trustworthy model forecasts and risk assessments in CCS projects.

Attempting to decrease the cost of CCS operation, CO_2 injection with impurity gas, i.e., N_2 and O_2, into a depleted gas reservoir was investigated. The impacts of the key parameters on the performance of CO_2 storage and CSEGR were analyzed in detail. The results showed that the effect of impurities on CO_2 storage capacity is dependent on the reservoir pressure and temperature conditions, and the concentration of impurities. The depleted gas reservoir with a relatively low temperature and low irreducible water saturation is favorable to the CO_2 storage capacity. A low primary gas recovery for the depleted gas reservoir is in favor of CSEGR, while it is suitable for dedicated CO_2 storage when the primary gas recovery is high. In addition, it is suggested to produce the CH_4 as possible before the operation of CO_2 storage and CSEGR. The chromatographic partitioning phenomenon may occur when N_2 and O_2 were co-injected with CO_2 into depleted gas reservoirs, which could be used as a monitoring strategy for the CO_2 front and potential CO_2 leakage. In addition to the solubility and concentration of the impurity gas would affect this phenomenon, there is a critical water saturation for the occurrence of significant chromatographic partitioning phenomenon associated with determined type and concentration of impurity gas.

To increase the cost-effectiveness of CCS, the suitability of utilizing CO_2 as the cushion gas in the UGSR was analyzed based on the geological parameters of Donghae depleted gas reservoir in Korea. The cyclic CH_4 production and injection were conducted over a period of 15 years to acquire the mixing behavior of CO_2 and CH_4 in a relatively long-term period. The results showed that the maximum CO_2 concentration that can be used for cushion gas is 9% under the condition of production and injection for 120 and 180 days in a production cycle at a rate of 4.05 and 2.7 kg/s, respectively. The typical curve of the mixing zone thickness can be divided into four stages, i.e., the increasing stage, smooth stage, suddenly increasing stage, and periodic change stage. The CO_2 fraction in the UGSR, reservoir permeability, and production rate have a significant effect on the breakthrough of CO_2 in the production well, while the effect of water saturation and temperature is neglectable. For the purpose of utilizing more CO_2 as cushion gas in the UGSR, CO_2 is supposed to be injected for supplementation during the operation of UGSR.

Generally, the parametric uncertainty analysis conducted in this thesis is beneficial for the risk assessments in CCS projects. Co-injecting CO_2 with impurities associated with CSEGR and utilizing CO_2 as cushion gas in UGSR are favorable for improving the economic incentives of CCS operation. Therefore, this thesis is beneficial for promoting the application of CCS and mitigating the atmospheric CO_2 emissions.

List of contents

List of figures

List of tables

1 Introduction

In this chapter, the background and the latest developments of CO_2 storage is introduced. Specifically, the background and the mechanisms and strategies of CO_2 storage, focusing on their characteristics and current status, are presented firstly. Then the strategies for assessing and ensuring the security of CO_2 storage operations, including the risks assessment approach and monitoring technology associated with CO_2 storage, are outlined. In addition, the engineering methods to accelerate CO_2 dissolution and mineral carbonation for fixing the mobile CO_2 are also compared. Further, the strategies for improving economics of CO_2 storage operations, namely enhanced industrial production with CO_2 storage to generate additional profit, and co-injection of CO_2 with impurities to reduce the cost are discussed. Based on the literature review, this thesis aims at reduce the risks related to CCS and increases the cost-effectiveness of CCS. The research objectives and outline of this thesis are also presented. The main contents of this chapter have been published in the following research paper (Cao et al. 2020): A review of CO_2 storage in view of safety and cost-effectiveness. *Energies*, 13(3), 600.

1.1 Introduction of underground CO_2 storage

The CO_2 concentration in the atmosphere locates at a level of below 300 ppm in pre-industrial times, whereas it has already risen above 410 ppm in the last few centuries (Met Office 2017; Scripps CO_2 Program 2019). Especially, the CO_2 concentration increases dramatically since 1960s as can be seen in Fig. 1.1. Fig. 1.1 also shows the correlation between the atmospheric concentration of CO_2 and the global temperature since 1850s. It can be seen that the continuous rise in global temperature is strongly related to the atmospheric concentration of CO_2, which indicates that CO_2 is the main contributor to global warming and climate change. More importantly, it is estimated that CO_2 makes up an 77% of greenhouse gases across the world (MacDowell et al. 2010; Rahman et al. 2017). Furthermore, the CO_2 emission may increase the frequency of extreme weather such as the extreme extratropical cyclones. Specifically, it is estimated that the number of extratropical cyclones will be more than triple by the end of this century in North America and Europe if the greenhouse gas emissions hasn't been efficiently mitigated (Hawcroft et al. 2018). To deal with such intense global climate problem, the Intergovernmental Panel on Climate Change's (IPCC) suggested that the increment of the average earth's surface temperature should be limited less than 2 °C within this century based on the Integrated Assessment Models (IAMs)'s estimation (Edenhofer et al. 2014).

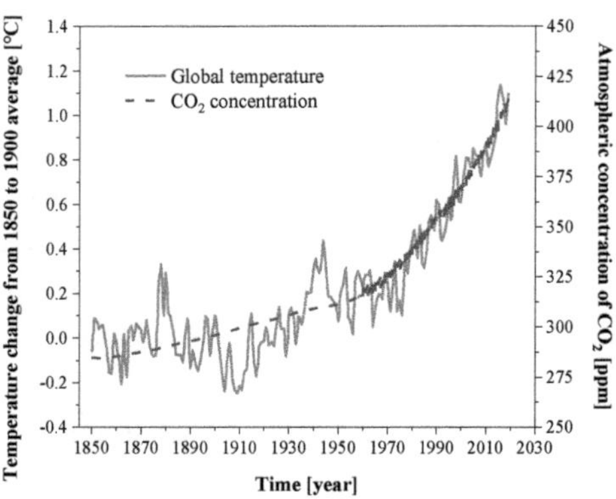

Figure 1.1 Correlation between atmospheric concentration of CO_2 and the global temperature since 1850s (Cao et al. 2020)

To achieve the IPCC's goal on global temperature control, carbon capture and storage (CCS) is supposed to be promoted. This is result from that CCS is currently regarded as the most effective strategy for slowing down the atmospheric CO_2 emissions and attenuating associated climate problems (Brinckerhoff 2011). As can be seen in Fig. 1.2, it is estimated that approximately 10.8 Gt CO_2 can be trapped through CCS alone by 2050, which undertakes almost 19% reduction in global CO_2 emissions (IEA 2010). Further, the overall cost of achieving the same targets of CO_2 emission reduction will increase by 70% without the application of CCS (IEA 2009), demonstrating the importance of CCS on the mitigation of atmospheric CO_2 emissions from the economic point of view as well. It should be mentioned that CCS is also beneficial for the circulation carbon economy, which offers a realistic and technology-neutral strategy that focusses on carbon management and will ultimately lead to a carbon-neutral energy future (IEF 2020).

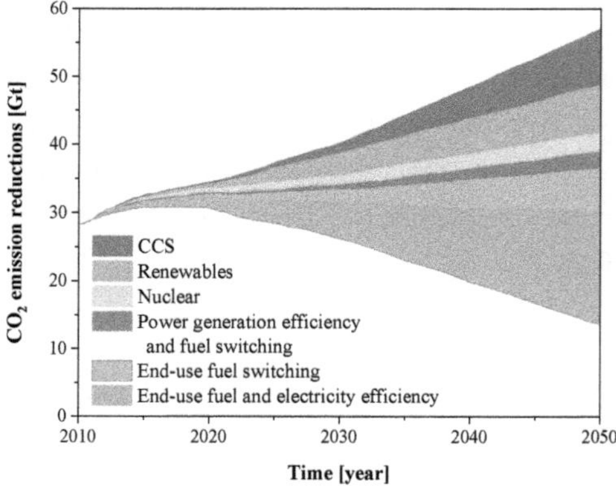

Figure 1.2 IEA forecasts of key technologies for CO_2 emission reductions (Cao et al. 2020; IEA 2010)

As can be seen in Fig. 1.3, a total of 51 CCS engineering projects are projected across the world, which are mainly scheduled in North America, Australia, Western Europe, and China. It should be mentioned that only 19 CCS projects are currently in operation (Global CCS Institute 2019). The main factors challenge the large-scale application of CCS are the high cost and safety risk associated with CO_2 leakage, even though CCS has been proven to be technically feasible. As a result, the contribution of CCS is still very limited in mitigating climate change (Gislason and Oelkers 2014; Pawar et al. 2015). Therefore, more research efforts on improving the safety and economics of CCS are required to develop this kind of technology, improving public acceptance, gaining support from government, and to accelerate the application of CCS in large-scale.

Figure 1.3 Commercial-scale integrated CCS projects around the world. Circle size is proportional to the CO_2 capture capacity and the color indicates different stages of the lifecycle of the project (Cao et al. 2020)

The review literatures in the past ten years on CCS technology are summarized in Tab. 1.1. It can be seen that almost every aspect of CCS technology including CO_2 capture and utilization, options for CO_2 storage and CCS projects, CO_2-brine-rock systems, well integrity and risk assessment, and storage efficiency and environmental considerations have been discussed extensively in the last decade (Abid et al. 2015; Abidoye et al. 2015; Aminu et al. 2017; Atia and Mohammedi 2018; Bachu 2015; Bai et al. 2016; Boot-Handford et al. 2014; Burnside and Naylor 2014; Carroll et al. 2014; De Silva et al. 2015; Godec et al. 2014; Kemper 2015; Koytsoumpa et al. 2018; Li et al. 2013; Li and Liu 2016; Liu et al. 2017; Mayer et al. 2015; Michael et al. 2010; Oh 2010; Pan et al. 2016; Pires et al. 2011; Riaz and Cinar 2014; Sanna et al. 2014; Shukla et al. 2010; Singh and Haines 2014; Song and Zhang 2013; Tan et al. 2016; Tang et al. 2014; Verduyn et al. 2011; Wee 2013; Zahid et al. 2011; Zhang and Bachu 2011).

However, the strategies for improving the safety and economics of CCS have not been discussed in detail. In addition, the technology of CCS is developing rapidly so that the recent development needs to be reviewed and discussed.

Table 1.1 Summary of review literature on CCS technology (Cao et al. 2020)

Research fields	Source	Review scope
CO_2 capture and utilization	Atia and Mohammedi 2018	Review of the application of CO_2 for enhanced oil and gas recovery
	Koytsoumpa et al. 2018	Review of CO_2 capture and reuse technologies, highlighting the strategies of CO_2 capture in variety of scenarios, and the state of the art for CO_2 utilization
	Li et al. 2013	Review of CO_2 capture, utilization, and storage (CCUS) in Chinese Academy of Sciences, highlighting the strategies for CCUS in China
	Tan et al. 2016	Review of the property impacts of CCS, highlighting the effect of uncertainties in thermal-physical properties on the design of components and processes in CCS
	Boot-Handford et al. 2014	Review of CCS highlighting the CO_2 capture technologies, the pilot plants, and the economic and legal aspects of CCS
	Godec et al. 2014	Review of CO_2 enhanced coalbed methane recovery, highlighting the CO_2 storage trials in the San Juan Basin in USA, and the estimation of CO_2 storage capacity in coal seams
	Liu et al. 2017	Review of CCUS technologies highlighting the engineering projects and their developments in China
	Pires et al. 2011	Review of CCS highlighting the findings obtained in CCS operational projects including the technologies of CO_2 capture, separate, transport, and storage
Options for CO_2 storage and CCS projects	Aminu et al. 2017	Review of CCS highlighting the options for CO_2 storage, the evaluation criteria for CO_2 storage site, and the major CO_2 storage projects
	Kemper 2015	Review of biomass with CCS (Bio-CCS), highlighting the economics and global status of Bio-CCS, and the role of Bio-CCS in the food-water-energy-climate nexus
	Michael et al. 2010	Review of CO_2 storage in saline aquifers, highlighting the geological and operation parameters, and the monitoring technologies for existing saline aquifers storage operations
	Oh 2010	Review of the CCS in coal-fired plant in Malaysia, highlighting the choices of coal plants and the capture technologies
	Riaz and Cinar 2014	Review of CO_2 storage in saline formations, highlighting the modeling of solubility trapping
	Sanna et al. 2014	Review of mineral carbonation (MC) technologies for CO_2 sequestration, highlighting the mechanisms of MC technologies and their contribution in decreasing the cost of CCS
	Singh and Haines 2014	Review of CCS projects and future opportunities, highlighting the technical details and business plan for CCS projects
	Tang et al. 2014	Review of CO_2 storage projects in China, highlighting the CO_2 source, and CO_2 storage strategies in China

	Verduyn et al. 2011	Review of CO_2 mineralization product forms, highlighting the mineralization process for CO_2 storage
	Wee 2013	Review of CCS by using coal fly ash, highlighting the feasibility and prospects of CCS using coal fly ash
CO_2-brine-rock systems	Burnside and Naylor 2014	Review of the relative permeability and residual trapping in CO_2 storage systems, highlighting the estimating and measuring methods
	De Silva et al. 2015	Review of the geochemical aspects of CO_2 storage in saline aquifers, highlighting the advantages of CO_2 storage in saline aquifers, and the CO_2-brine-rock interactions in the aquifers
	Pan et al. 2016	Review of geomechanical modeling of CO_2 storage, highlighting the numerical methods and their application in the modeling of ground deformation, faults, and fracture propagation
	Abidoye et al. 2015	Review of CO_2 sequestration highlighting the trapping mechanisms and the flow of CO_2-brine in porous media system
Well integrity and risk assessment	Abid et al. 2015	Review of the cement degration in CO_2-rich condition of CCS projects highlighting the degration of Portland cement
	Li and Liu 2016	Review of the risk assessment of CO_2 storage, highlighting the regulations and strategies of risk assessment for CO_2 storage
	Mayer et al. 2015	Review of the isotopic composition of CO_2 for leakage monitoring in CCS project, highlighting the stable isotopes as a tracer for injected CO_2
	Zhang and Bachu 2011	Review of the integrity of existing wells for CCS, highlighting the mechanical well failure and chemical issue due to cement carbonation
	Bai et al. 2016	Review of well integrity of CCS highlighting the corrosion of metallic and cement, and the remedial measures
	Song and Zhang 2013	Review of caprock sealing mechanisms for CO_2 storage, highlighting the problems associated with CO_2 leakage, the leakage paths, and the factors that affect leakage
	Zahid et al. 2011	Review of CO_2 storage highlighting the capacity estimation of storage sites, the monitoring technologies and simulation tools for CCS
	Shukla et al. 2010	Review of CO_2 storage and caprock integrity, highlighting the major CCS project in operation and CO2 migration in the reservoirs
Storage efficiency and environmental considerations	Bachu 2015	Review of CO_2 storage efficiency in saline aquifers, highlighting the factors that affect CO_2 plume migration and the methods to estimate the storage capacity
	Carroll et al. 2014	Review of environmental considerations for CO_2 storage in sub-seabed, highlighting the potential ecological impacts

In the following section, the most recent progress on addressing the challenges related to assessing and decreasing the risks of CO_2 leakage, cutting the cost of CO_2 storage, and promoting the developments of commercial scale CCS projects will be reviewed and analyzed. Firstly, the mechanisms of CO_2 storage and the strategies of CO_2 storage are reviewed and discussed. Then the risk assessment of CO_2 storage and strategies for decreasing the risks of CO_2 leakage, including accelerating CO_2 dissolution and mineral carbonation, are summarized. Finally, the strategies for cutting the cost and acquiring additional benefits of CO_2 storage to improve its cost-effectiveness, including co-injection of CO_2 with impurities and enhanced industrial production with CO_2 storage, are discussed.

1.2 Mechanisms of CO_2 storage

Figure 1.4 shows the phase diagram of CO_2. Considering that the pressure and temperature in the process of CO_2 storage is range from approximately 5 to 60 MPa and 20 to 150 °C respectively, thus the CO_2 may in the gaseous and supercritical state. For instance, when the pressure and temperature reach to the critical pressure and critical temperature, i.e., 7.38 MPa and 31.04 °C, CO_2 will exit in supercritical state and owns the characters of both gaseous CO_2 and liquid CO_2. On the one hand, the supercritical CO_2 has a low viscosity like gas, which is beneficial for improving the injectivity. On the other hand, the supercritical CO_2 has a high density like liquid, which is beneficial for improving the storage capacity in CCS systems.

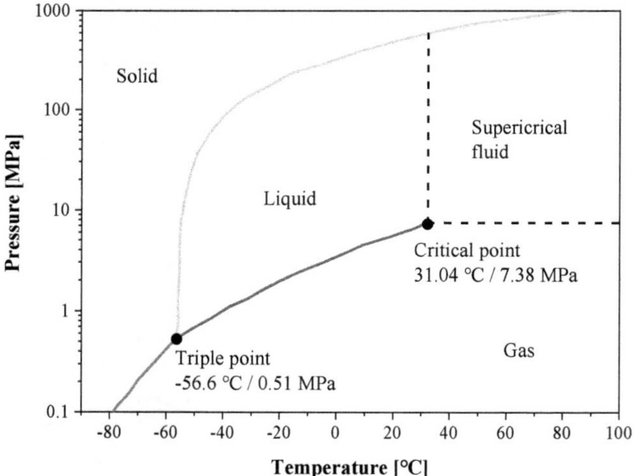

Figure 1.4 Phase diagram of fluid carbon dioxide (Data from Vargaftik 1975)

After the CO_2 has been injected into underground reservoirs, four main CO_2 trapping mechanisms may play a role on the trapping of CO_2 storage. As shown in Fig. 1.5a, the CO_2 trapping mechanisms consist of structural and stratigraphic trapping, residual trapping, solubility trapping, and mineral trapping (Shukla et al. 2010). The structural and stratigraphic trapping is regarded as the most dominant trapping mechanism. Once CO_2 is injected into subsurface reservoir formations, it will migrate upward to the top of geological structures owing to the buoyancy effect. Then the CO_2 will stay below the impermeable caprock. Regarding the residual trapping, the injected CO_2 will displace formation fluids when it migrates through the reservoir rock. Further, the displaced fluid disconnects and traps the remaining CO_2 within the pores of rocks due to the capillary force (Bradshaw et al. 2007). In the residual trapping, the CO_2 is trapped by capillary force. It can achieve trapped CO_2 at a saturation of at least 10% and even reach more than 30% of the pore volume in some formation rocks (Krevor et al. 2015; Zhang and Huisingh 2017). Regarding solubility trapping, CO_2 will dissolve in formation fluids and become immobile, thus decreasing the mole fraction of free CO_2 (Bian et al. 2019). It should be mentioned that the dissolved CO_2 will slightly increase the density of formation fluids by around 1%, which is sufficient to promote the convection flow with the help of such a small density difference (Zhang et al. 2008). This convection flowing is also in favor of the trapping of CO_2. Under the temperature, pressure, and salinity conditions of conventional CCS reservoirs, the solubility of CO_2 in groundwater ranges from 2% to 6%. It should be pointed that the solubility of CO_2 decreases with the growing temperature and salinity (Zhang and Huisingh 2017). In mineral trapping mechanism, CO_2 is trapped by the geochemical reactions with the rocks in reservoir. The CO_2 usually precipitates as carbonate so that it can be trapped in immobile secondary phases effectively (Sundal et al. 2014).

As shown in Fig. 1.5b, different trapping mechanism plays different role on CO_2 storage in the time scale between 1 and 10,000 years. It can be seen that the structural trapping plays an important role in the initial stage of CO_2 storage. However, the effect of structural trapping becomes weak gradually. Fig. 1.5a also shows that the residual trapping and solubility trapping have a significant impact in the time scale of tens of years. Further, the residual trapping and solubility trapping would lock up a certain amount of CO_2 for thousands of years. Regarding the mineral trapping, it begins to work at almost around one hundred years and its effect would increases gradually. Finally, the mineral trapping can play a key role in a geological timescale.

(a)

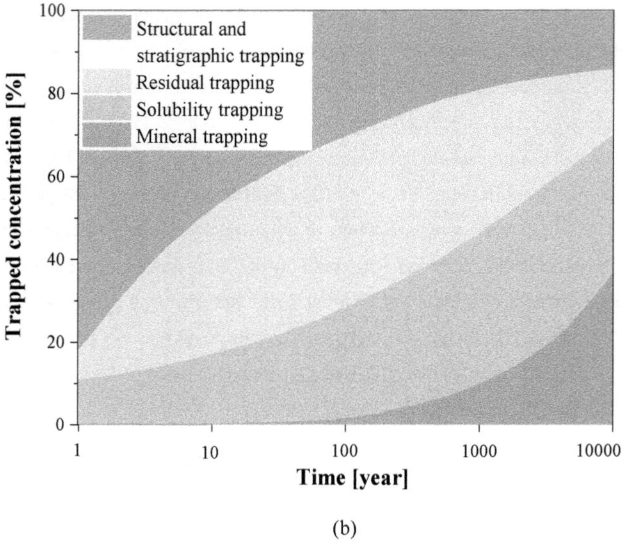

(b)

Figure 1.5 (a) The four main CO_2 trapping mechanisms (Zhao et al. 2014); (b) the contribution of four CO_2 trapping mechanisms with time (Cao et al. 2020; Metz et al. 2005)

1.3 Geologic storage options of CO_2

1.3.1 Saline aquifers

CO_2 storage in saline aquifers is one of the most important strategies because of the huge amount of storage capacity. It is estimated that approximately 10,000 Gt of CO_2 could be sequestrated by the saline aquifers around the world. In other words, the saline aquifers are sufficiently store the CO_2 emissions from large stationary sources for more than 100 years (Celia et al. 2015; Davison et al. 2001; De Silva et al. 2015). Furthermore, the saline aquifers usually have a greater regional coverage and more wide distribution compared with the other storage options. Therefore, the saline aquifers have a better chance to be located nearby the sources of CO_2 emission, which could reduce the cost of CO_2 transportation (Cooper 2009; Zhang and Huisingh 2017). There are two crucial problem brought by CO_2 storage in saline aquifers. The first one is the pressure build up, which has the potential to lead to the fracturing of formation and the reactivation of faults. The second one is the CO_2 plume migration in formation, which may lead to the leakage of CO_2 that should be paid more attention (Orlic 2016). Birkholzer et al. (2009) conducted a numerical simulation to investigate the impact of large-scale CO_2 sequestration with an injection rate of 1.52 million tons per year (Mtpa) in a saline aquifer open boundary. The results showed that there is significant pressure build up in the reservoir formation at the zone even more than 100 km away from the injection zone, whereas the CO_2 plume migration is rather small that is approximately 2 km and is concentrated on the top of saline aquifer caused by the buoyancy effect. Their results also showed that the pressure perturbation could affect the shallow groundwater formation if there is a caprock with relatively high permeability (higher than 10^{-18} m^2) between the shallow layers and the saline aquifer. Fortunately, it should be mentioned that the migration of reservoir fluids, i.e., the CO_2 and formation water, into groundwater formation is extremely unlikely. This demonstrates the safety and suitability of large-scale CO_2 sequestration in saline aquifers.

A total of five commercial-scale CCS projects across the world have been launched in saline aquifers, including the Sleipner project (Audigane et al. 2007; Audigane et al. 2006; Williams and Chadwick 2018), the Snøhvit project (Hansen et al. 2013), the In Salah project (Ringrose et al. 2013; Rutqvist et al. 2010), the Gorgon project (Flett et al. 2008), and the Quest project (Bourne et al. 2014). Regarding the Sleipner project, the CO_2 was injected into a saline aquifer within the Utsira Sand formation. The injected CO_2 was separated from the produced natural gas at the Sleipner field in the North Sea. Generally, a total of 18 million tons of CO_2 has been injected by 2018 since the initiation in 1996 (Williams and Chadwick 2018). Based on the engineering experiences of the Sleipner project, the Snøhvit CCS project that is located in the Barents Sea was launched in 2008 with a total amount of 1600 ktons of CO_2 injected till August 2012. In this project, the CO_2 separated from the LNG project was injected into the deeper Tubåen Formation. It is scheduled that around 23 million tons of CO_2 would be sequestrated in the reservoir based on the projected lifetime of the Snøhvit LNG project (Hansen et al. 2013; Simmenes et al. 2013).

The project located at In Salah, Algeria, is a pioneering CCS project across the world. A total of more than 3.8 million tons of CO_2 have been injected into the Krechba field since 2004 (Ringrose et al. 2013). It's worth to be mentioned that the diversity of monitoring methods including satellite monitoring and 4D seismic have been used in this CCS project to monitor the response of formation to CO_2 injection. Meanwhile, the accessibility of the monitoring data to the public is very high (Bjørnarå et al. 2018; Eiken et al., 2011; Gemmer et al. 2012; Newell et al. 2017; Rinaldi and Rutqvist 2013; Rinaldi et al. 2017; Ringrose et al. 2013; Rutqvist et al. 2010; Shi et al. 2013; Stork et al. 2015), so it could be served as a commendable case to investigate the CCS in saline aquifers.

The Quest CCS project launched in 2015, which is designed to store the CO_2 from an existing facility for upgrading heavy oil in Scotford of Alberta, Canada. It is expected that around 27 million tons of CO_2 could be injected into the Basal Cambrian Sands formation through 3 to 8 vertical wells with an injection rate of 1.08 Mtpa (Bourne et al., 2014).

The Gorgon CCS project is located in the northwest of Australia. There is a Jurassic saline reservoir in the Dupuy Formation that can be served as reservoirs for CO_2 storage. During the lifetime of the Gorgon project, a total of more than 120 million tons of CO_2 is planned to be injected into the Dupuy Formation at an injection rate of 3.8 Mtpa (Flett et al. 2008).

Aside from the forementioned large-scale CCS projects, there are some small-scale projects as well, including the Illinois Basin-Decatur Project (Finley et al. 2013), Ketzin pilot site (Martens et al. 2012; Opedal 2018), and Shenhua CCS demonstration project (Yang et al. 2017). Generally, these CCS projects have been conducted with detailed modeling and monitoring during operation, which demonstrates the safety and suitability of this technology. At the same time, it helps increase the public acceptance about CCS technology.

However, although the CO_2 storage capacity of saline aquifers is huge, the overall application of CO_2 storage in saline aquifers across the world is still at a small-scale because of the lack of financial incentives. Therefore, the policies related to the taxes on carbon emission may need to be formulated, which demonstrates the important role on the application of CCS should be played by the government.

1.3.2 Depleted oil and gas reservoirs

There are many merits for CO_2 storage in depleted oil and gas reservoirs. Firstly, there are many existing equipment installed on the surface and underground in depleted oil and gas reservoirs, thus it can be reused for CO_2 sequestration with only minor modification. Secondly, the seal quality and the integrity of the caprock are guaranteed. The geological conditions of the depleted oil and gas reservoirs have also been comprehensively characterized during the exploration and production process (Orlic 2016). Thirdly, the change of induced stress and the extent of pressure perturbations is much smaller compared with saline aquifers due to the long-term extraction of oil and gas from the reservoirs (Orlic 2016). It should be mentioned that the depleted gas reservoirs are more favorable for CCS compared with depleted oil reservoirs. This is result from that a larger CO_2 storage capacity per pore volume is available due to the higher compressibility of gas and ultimate recovery (Barrufet et al. 2010; Mamora and Seo 2002; Stein et al. 2010). Regarding the types of gas reservoirs used in this form of storage, the condensate gas reservoirs are more advantageous over the wet and dry gas reservoirs. There are several reasons account for it. Firstly, there is little gas remained in the condensate gas reservoirs thus more effective volume could be used for CO_2 sequestration. Secondly, the phase behavior of the mixture of condensate gas and CO_2 is favourable for CO_2 sequestration. Thirdly, the good gas injectivity is accompanied with the condensate gas reservoirs (Raza et al. 2018). Furthermore, the stored CO_2 per pore volume in depleted condensate reservoirs is very high. Specifically, it is approximately 13 times higher than that of the equivalent aquifer (Barrufet et al. 2010). However, it should be mentioned that the phase change may occur in depleted condensate reservoirs that should be paid for attention.

There are some characteristics associated with the long-term trapping mechanisms of CO_2 in natural gas fields. It is reported that the solubility trapping in formation water is dominated while the mineral trapping is limited in the natural gas reservoirs with siliciclastic or carbonate lithologies. This is verified by the results of noble gas and carbon isotope traces (Gilfillan et al. 2009). It is worth to mention that the residual gas saturation in the depleted reservoirs has an impact on the CO_2 storage capacity.

Specifically, the capillary trapping capacity usually exhibits a positive correlation with the remaining gas saturation, while the dissolution trapping capacity, structural trapping capacity, and the total storage capacity are inversely related with it (Raza et al. 2018).

It should be mentioned that there may be some problems related to CO_2 storage in depleted gas reservoirs such as the Joule-Thomson cooling effect. The Joule-Thomson cooling effect may occur in the reservoirs with low pressure at the initial stage of injection, which may lead to strong reduction of the reservoir temperature, further forming hydrate, freezing the residual water, and even compromising the well injectivity, especially when CO_2 is injected with low temperature (Mathias et al. 2010; Oldenburg 2007; Twerda et al. 2018). The Joule-Thomson cooling may lead to the formation of hydrate in the reservoir with a reservoir temperature of less than 20 °C through the initial reservoir pressure reach to 6 MPa. However, the Joule-Thomson cooling effect is not noticeable in permeable reservoirs when the reservoir temperature is over 40 °C, even though the formation pressure is as low as 2 MPa (Mathias et al. 2010). In order to avoid the Joule-Thomson cooling, CO_2 are supposed to be injected with high temperature or a high mass flow rate. It should be mentioned that the high mass flow rate may cause some other problems at the beginning and shut-in of the gas injection. The system in ROAD project connects a CO_2 capture system at Masvlakte Power Plant with a depleted gas field. The depleted pressure of the reservoir is less than 2 MPa (Böser and Belfroid 2013). There is a single source and sink system that allows the control of pressure and temperature at the shoreline inlet of the offshore pipeline by adjusting the level of after cooling at the compressor. It can achieve a high downhole temperature so that ease the Joule-Thomson cooling effect. The working mechanism is that injecting CO_2 with high temperature into the reservoir with low pressure, whereas injecting CO_2 with low temperature in the reservoir with higher pressure. In this way, the injection pressure requirement would be maintained at a low level (Twerda et al. 2018). It should be noted that the co-injection of SO_2 and CO_2 is an alternative method to reduce the Joule-Thomson cooling effect result from the beneficial thermal consequence (Ziabakhsh-Ganji and Kooi 2014). In addition, it is reported that the presence of methane can potentially suppress the Joule-Thomson cooling effect (Loeve et al. 2014).

Though the technology of CO_2 storage in depleted gas reservoirs has been identified technical feasible, only few field projects dedicated to CO_2 storage have been implemented. The first demonstration project in Australia is the CO2CRC Otway Project (Sharma et al. 2009). In this project, CO_2 was injected into the Waarre C Formation at a depth of around 2050 m. It was commenced in March 2008 and ended in August 2009, with a total CO_2 storage capacity of 65,445 tons (Jenkins et al. 2012). It is worth to mention that a community led "stakeholder reference group" has been set up in this project to provide a channel for communicating with the public, which is beneficial for increasing their acceptance about CCS technology. This could be served as a demonstration for other CCS projects. Overall, the CO2CRC Otway Project demonstrates that CO_2 storage in depleted gas fields can be achieved safely and efficiently (Jenkins et al. 2012). It also lays a foundation for the large-scale CO_2 storage in depleted oil and gas fields. According to the experience obtained in this project, the suitability of CO_2 storage in some other depleted gas reservoirs has been evaluated. For example, the depleted P18-4 gas field on the offshore of Netherlands (Arts et al. 2012) and the DF-1 South China Sea Gas field (Zhang et al. 2010), have been identified as suitable sites for CO_2 storage.

Generally, before the wide application of large-scale CCS in saline aquifers, CCS in depleted reservoirs can make a great contribution in the mitigation of global warming because of its advantages of low risk and cost-effectiveness (Hannis et al. 2017).

1.3.3 Coal bed

CO_2 injection in coal beds is an attractive strategy for CO_2 storage. It is reported that most of the suitable coal beds for the sequestration of CO_2 are located at a depth ranging from 300 to 900 m (Bachu 2007). The major advantage for CO_2 storage in deep ocean is that the potential coal beds are usually located nearby the existing or planned coal-fired power plants, thus the transportation cost could be reduced significantly. However, CO_2 storage in coal beds is currently an immature technology. There are only some pilot studies conducted on its suitability and storage capacity. For example, the evaluated effective storage capacity of Cretaceous-Tertiary coal beds in Alberta, Canada is 6.4 Gt (Bachu 2007), and the potential storage capacity for the coal beds in China is around 142.67 Gt (Yu et al. 2007). These studies signify the potential contribution of CO_2 storage in coal beds on the mitigation of CO_2 emissions.

1.3.4 Deep ocean

CO_2 can also be directly injected into the deep ocean when the water depth is more than 2,700 m (Brewer et al. 1999; Fer and Haugan 2003). In this case, the liquid CO_2 can sink downward to the seafloor result from that the density of CO_2 is higher than that of seawater under the high pressure and low temperature conditions (Fer and Haugan 2003; Levine et al. 2007). The storage capacity of this kind of technology is extremely huge due to the enormous volume of the ocean. However, CO_2 storage in deep ocean cannot be applied in large-scale because it may be harmful to the marine environment.

1.3.5 Deep-sea sediments

The technology of CO_2 storage in deep-sea sediments not only combines the advantages of ocean storage and geologic storage, but also avoids many shortcomings (House et al. 2006; Koide et al. 1997; Schrag 2009; Teng and Zhang 2018). For example, the ocean ecosystem is free from been polluted because the CO_2 is injected into the deep-sea sediments rather than directly into the ocean. Regarding the storage mechanisms, the terrestrial storage mechanisms including the residual trapping, dissolution trapping, and mineral trapping still play a positive role. Besides, several new storage mechanisms such as the gravitational trapping and hydrate trapping could also work in the storage. The gravitational trapping comes from the higher density of CO_2 that drives the CO_2 migrating downward into the deep sea to the so-called negative buoyancy zone (NBZ) (Levine et al. 2007). It should be mentioned that the depth at which the density of CO_2 is equals to that of seawater at the salinity and temperature conditions is around 2,700 m (Metz et al. 2005). The hydrate trapping works due to the formation of CO_2 hydrate under low temperature and high pressure conditions (House et al. 2006). The long-term evolution of injected CO_2 in the deep-sea sediments is shown in Fig. 1.6. It can be seen that the hydrates form at the bottom of hydrate formation zone (HFZ), which is beneficial for reducing the permeability of the caprock. The area of hydrate caprock expands gradually along with more CO_2 reaching the bottom of the HFZ. Meanwhile, the aqueous saturated with CO_2 would sink downward due to the advection effect that is driven buoyancy. Finally, the hydrate CO_2 and liquid CO_2 will dissolve in seawater and transform into CO_2 aqueous solution by diffusion effect, thus the permanent storage occurs.

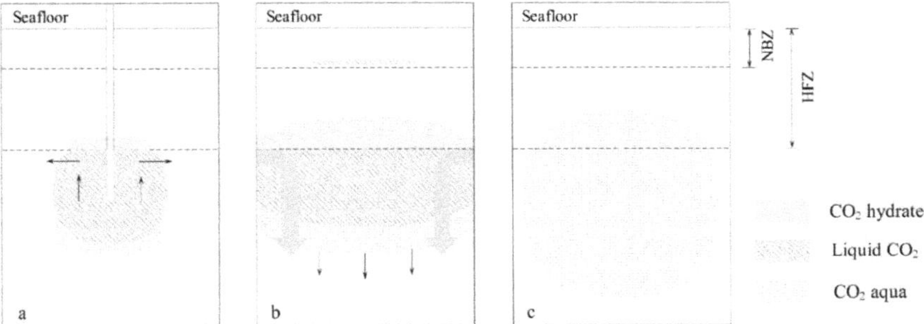

Figure 1.6 The long-term evolution of the injected CO_2 (Cao et al. 2020; House et al. 2006)

Though CO_2 storage in deep-sea sediments has been identified technological feasible, it is still in the formulation technology readiness level. It should be mentioned that CO_2 storage in deep-sea sediments is far more expensive than onshore storage. In addition, it may lead to some ecological problems such as the respiratory stress, acidosis, and metabolic depression for variety of organisms (Adams and Caldeira 2008; Schrag 2009). Therefore, a long time may be needed to increase the public acceptance of this method.

In a summary of this section, there are several strategies for the sequestration of CO_2, including CO_2 storage in saline aquifers, depleted oil and gas reservoirs, coal beds, deep ocean, and the deep-sea sediments. The pros and cons of these storage strategies are summarized in Tab. 1.2.

Table 1.2 Summary of the pros and cons of the CCS technologies (Cao et al. 2020)

Option	Pros	Cons
Saline aquifers	Huge amount of storage capacity, wide distribution, commercial technology readiness level	No economic benefit
Depleted oil and gas reservoirs	Existing installed equipment, guaranteed caprock integrity, characterized geological conditions, small pressure perturbations and induced stress changes, additional oil, and gas recovery	Demonstration technology readiness level
Coal beds	Low transportation cost due to it may located nearby the coal-fired power plants, additional coalbed methane recovery	Pilot plant technology readiness level
Deep ocean	Large storage capacity	Formulation technology readiness level, no economic benefit, may affect the marine environment
Deep-sea sediments	Enormous storage capacity, free from the potential harm to the ocean ecosystems	Formulation technology readiness level, no economic benefit, far more expensive than onshore methods

It can be clearly seen from Tab. 1.2 that only the technology of CCS in saline aquifers has been commercially used among these technologies. It is also regarded as a promising CCS strategy because of its huge amount of storage capacity. However, there is no economic incentives for this kind of storage technology, which would be a main obstacle for its large-scale application. Considering the relatively low cost and potential additional profit for CCS in deleted oil and gas reservoirs, it may become another promising CCS technology in the near future. It should be mentioned that the leakage risk of CO_2 along the abandoned wellbore is one of the most important factors restrict the application of CO_2 storage in depleted oil and gas reservoirs. Therefore, the long-term integrity of wellbore should be analyzed in detail. Specifically, the long-term experiments and molecular dynamic simulations should be conducted to study the kinetics between CO_2 with the cement, well string, as well as formation rocks under the related physical and chemical conditions.

1.4 Risk assessment of underground CO_2 storage

1.4.1 Simulation of underground CO_2 storage

There are some issues due to CO_2 injection should be taken into consideration. Firstly, the migration of CO_2 and formation fluids caused by gas injection may affect the ground water resources (Birkholzer et al. 2009). In addition, the chemical reaction between CO_2 with the cement and well string may lead to the failure of well integrity (Gaus et al. 2008; Gawel et al. 2017; Opedal et al. 2018). The formation of fracture, the reactivation of faults and the shear failure of caprock can also lead to the failure of caprock integrity, resulting the leakage of CO_2 (Pawar et al. 2015; Shukla et al. 2010). Therefore, it is very important to assess the related risks of CCS operation through predicting the formation responses of CO_2 injection, including the formation deformation, formation pressure change, and the migration of

CO_2 plume etc. Generally, such temporal and spatial responses of formation could be predicted by both analytical and numerical methods. However, due to the geological complexities of the formation and the physical complexities in CCS projects, only a few semi-analytical models have been developed to estimate the migration of CO_2 plume and pressure distribution during CO_2 injection (Bao et al. 2014; Michael et al. 2011; Nordbotten et al. 2005; Wang and Wang 2018; Xu et al. 2012). For example, Nordbotten et al. (Nordbotten et al. 2005) derived the solution for the evolution of CO_2 plume during injection in dimensionless form as follows:

$$\begin{aligned}&\left(Q_{\mathrm{well}}\left(p\left(r_{\mathrm{well}},t\right)-p_{\mathrm{init}}\right)+\Delta E_{\mathrm{p}}\right)\left(\frac{2\pi\lambda_{\mathrm{w}}kB}{Q_{\mathrm{well}}^{2}}\right)\\&=-\int_{0}^{1}\frac{(\lambda-1)\ln r'\left(b'\right)}{\left((\lambda-1)b'+1\right)^{2}}\mathrm{d}b'+\Gamma\int_{0}^{1}b'\left[r'\left(b'\right)\right]^{2}\mathrm{d}b'+\frac{1}{2}\frac{\lambda-1}{\lambda}\ln\left(\frac{\pi B\varphi}{V(t)}\right)\end{aligned} \tag{1.1}$$

Where Q_{well} denotes the volumetric injection rate; p_{init} denotes the initial pressure distribution; $p(r_{\mathrm{well}}, t)$ represents the pressure at the injection well; ΔE_{p} represents the potential energy as the energy required to submerge the lighter CO_2 into the denser water; B represents the thickness of the reservoir; λ_{w} denotes the mobility of water; λ is the total mobility; φ represents the porosity of the medium; k is intrinsic permeability; $V(t)$ is the total volume injected; b denotes the thickness in the CO_2 plume profile, where radial symmetry is assumed and the function b is taken to be a function of radial distance from the injection well (r) and time (t).

Due to the efficiency, the numerical methods are more popular in the evaluation of the risks associated with CO_2 storage. Many thermal-hydraulic (TH) coupled simulators have been developed and used for the computational simulation of multi-component and multi-phase flow in CO_2 storage, such as TOUGH (Birkholzer et al. 2009; Lengler et al. 2010; Wasch et al. 2013), MUFTE-UG (Ebigbo et al. 2006), ECLIPSE (Iogna et al. 2017; Rafiee and Ramazanian 2011), COMET3 (Schepers et al. 2009), CMG-GEM (Jia et al. 2016; Mishra et al. 2017; Ren 2018; Wriedt et al. 2014), STOMP (Bao et al. 2013; Hou et al. 2014), MRST (Allen et al. 2018), Tempest (Khan et al. 2012, 2013), IPARS (Jung et al. 2018), FLUENT (Luo et al. 2013). The thermal-hydrological-mechanical (THM) coupled simulators usually are constructed based on the coupling framework of fluid flow simulator (TH) and mechanical simulator (M). Specifically, the THM simulators used in the simulation of CCS mainly include TOUGH-FLAC3D (Rinaldi and Rutqvist 2017; Rutqvist 2011; Rutqvist et al. 2010), TOUGH-RBSN (Liu et al. 2013), TOUGH2-RDCA (Pan et al. 2014), Sierra Arpeggio (Aria-Adagio) (Newell et al. 2017), ABAQUS-ECLIPSE (Fei et al. 2014), OpenGeoSys-ECLIPSE (Benisch et al. 2013), and ECLIPSE-VISAGE (Shi et al. 2013). Among them, the TOUGH-FLAC has been verified and applied in CCS by many researchers (Gou et al. 2014; Rutqvist et al. 2010). It was developed by the Lawrence Berkeley National Laboratory (Rutqvist and Tsang 2002, 2003).

In recent years, some integrated software has been developed to assess the risks of CCS, such as the Leakage Assessment and Cost Estimation (PyLACE) and the National Risk Assessment Partnership (NRAP) Toolset. The NRAP is designed by U.S. Department of Energy's National Energy Technology Laboratory (Pawar et al. 2016). It can be used to evaluate the environmental risks associated with CCS operation. Fig. 1.7 shows the framework of NARP. Firstly, the geological system of CCS is divided into several discrete subsystems. The subsystem is then characterized with a reduced-order model (Argha et al. 2017). Further, the reduced-order models are linked together by an integrated assessment model based on the system modeling approach (Pawar et al. 2016). Finally, the whole system model can be used to assess the risk performance related to the CCS operation.

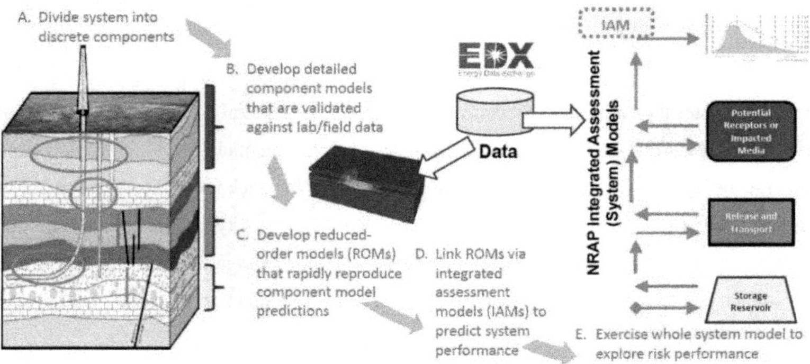

Figure 1.7 The framework of NARP (NETL 2017)

By using NARP, the risks related to CO_2 leakage and induced seismicity can be evaluated. Furthermore, the behavior of the components in CCS systems, including the wells, caprocks, reservoirs, and ground water aquifers, can be modeled by using corresponding tools. For example, the Wellbore Leakage Analysis Tool (WLAT) can be used to evaluate the leakage potential of existing wells (Doherty et al. 2017), and the Design for Risk Evaluation and Monitoring (DREAM) could be used for assessing and optimizing the designs of monitoring for long-term CO_2 storage operation.

The web application PyLACE based on Python is designed for quantifying the financial risks associated with potential CO_2 leakage in a CCS system (Sun et al. 2018). There are two major functional blocks in the PyLACE, the first one is metamodel development and the second one is metamodel-based decision support. It can convert the process-level risk assessment models into high-fidelity metamodels. Further, the online assessment can be conducted through the high-performance computing and cloud computing infrastructures.

Recently, the method of deep neural network inversion has been applied on 4D seismic data to estimate the saturation and pressure (Dramsch et al. 2019), proving the availability of deep neutral network on the application of data-based inversion. To make the evaluation of CCS more effective and efficient, the machine learning technology is encouraged to be used. It is forecasted that the assessment of CCS will be more and more intelligent with the development and application of machine learning technology.

1.4.2 Monitoring technologies in the assessment of CCS risks

The injected CO_2 will be retained in underground formation for a long time, and the CO_2 plume may have a harmful effect on the surrounding environment and the groundwater (Leung et al. 2014). It is difficult to assess the key issues or risks related to CCS with reasonable accuracy by utilizing only simulation tools (Nordbotten et al. 2012), thus the technology of monitoring and history matching is very important in the evaluation of CCS. The monitoring technologies have been widely used in the field development plans and routine field operations (Ringrose et al. 2013). The most used monitoring technology in CCS includes 3D seismic, micro-seismic, microbiology, geochemical sampling, distributed temperature sensing technology, vertical seismic profiling, gravimetry, cross-hole electromagnetic, pressure and temperature monitoring, 4D seismic, soil and gas sampling analysis, core analysis, atmospheric monitoring, tracers, and satellite monitoring.

The 3D seismic can provide a tri-dimensional image of the CO_2 plume and the formation structures. It should be mentioned that the quality of the 3D seismic is affected by the specific medium. In off-shore monitoring, the 3D seismic monitoring data with high quality could be obtained. Additionally, the CO_2 bodies above 10^6 kg located at the depth of 1 to 2 km can be identified because of the enhanced penetration of seismic waves in water (Leung et al. 2014).

The technology of 4D seismic involves the repeating 3D seismic in time-lapse mode to image the CO_2 plume in the reservoir over time, which is useful for monitoring the CO_2 migration. It should be mentioned that the non-repeatable noise level in the data challenges reflecting the field data by the 4D signal with high accuracy. This is result from that the seismic imaging experiments is difficult to be repeated from one survey to the other due to the variations in the soil moisture content, the formation water properties, the sources-receiver positioning and geometry (Lumley 2010).

The micro-seismic activity levels are correlated with the CO_2 injection periods, thus it is useful to understand the process of subsurface CO_2 injection and migration. The mechanism will be briefly introduced. In this procedure, the one-dimensional array that consists of some three-component downhole geophones would be deployed in the vertical well. Afterwards, the geophones could detect the waveform data and transfer it to the digitizers for recorded (Oye et al. 2013).

Apart from conventional surface seismic acquisition, it should be noted that the buried geophone, hydrophone arrays, and the fiber optic cables are permanently installed in the vertical seismic profiling systems. This is for the purpose to achieve a long-term monitoring and obtain the details of the geological structure (Götz et al. 2018). The vertical seismic profiling has been successfully applied in the Ketzin pilot site and the MRCSP project.

The gravimetry testing can detect the variations of formation fluid density caused by CO_2 injection. Therefore, it can provide the information related to the location of CO_2. It should be mentioned that the testing result by the gravimetry testing is affected by the shape of CO_2 plume (Kabirzadeh et al. 2017; Leung et al. 2014).

The technology of cross-hole electromagnetic is a non-invasive method on the characterization of the subsurface physical and chemical properties. It is worth to mention that the cross-hole electromagnetic can also provide the information related to the location of CO_2. The mechanism is as follows. Firstly, the electrical conductivity before and after the CO_2 injection would be collected. Secondly, the electrical conductivity will be converted to the CO_2 saturation by using appropriate rock-physics models and inversion algorithms (Böhm et al. 2015; Carcione et al. 2012). It should be noted that the cross-hole electromagnetic can only be used in small-scale areas such as the area between wells (Böhm et al. 2015).

Monitoring the fluid pressure and the temperature is very useful and important. On the one hand, the fluid pressure is beneficial for the evaluation of the risks associated with the failure of the caprock integrity. On the other hand, the variation of temperature can be used for identifying the flow path of the injected CO_2. It should be pointed out that the wellhead pressure and temperature cannot provide enough information for characterizing the CO_2 injection process. This has been certified at the Ketzin pilot site (Liebscher et al. 2013). Therefore, the monitoring of downhole pressure and temperature are recommended in the CO_2 storage operation.

The geochemical sampling analysis can detect the chemical variations, including the natural variations in water chemistry and the drop of pH, which is important for establishing a useful baseline for groundwater hydrology (Boreham et al. 2011). In addition, it can provide the information of the variation of the mineral concentration that caused by the dissolution of carbonates and precipitation of anhydrite. It should be pointed that the chemical reaction is a very slow process in the sandstone reservoirs (Gaus 2010), which is difficult to be detected in a short-term storage period.

Soil and gas monitoring can provide the information of CO_2 concentration, which is beneficial for the definition and determination of the baseline before CO_2 injection (Leung et al. 2014). In addition, it can provide more data on natural CO_2 variations in different environments and associated seasonal fluctuations.

Tracers monitoring is regarded as a cost-effective method for monitoring the origin of CO_2 observations in the underground complex storage formation. The mechanism of traces monitoring is as follows. Co-injection of CO_2 with some specific compounds that could be detected even in a very small concentration, thus the trail of the injected CO_2 can be reflected by the traces. The traces can be used in the monitoring of CCS including the SF_6, SF_5CF_3, and the isotope ^{14}C (Jenkins et al. 2012; Matter et al. 2016).

The technology of atmospheric monitoring can detect the atmospheric concentration of CO_2 that may varies due to the leakage of CO_2 from the underground. This is beneficial for the identification of the anomalies above the natural base line (Etheridge et al. 2011). It should be noted that the reliability of atmospheric monitoring may be affected by the significant natural variation of atmospheric concentration of CO_2, which is induced by the soil respiration and the organic matter decomposition (Leung et al. 2014).

Microbiology monitoring can be conducted on the samples of reservoir rocks and fluids before the injection of CO_2, which could be defined as a baseline of CO_2. Similarly, the results of microbiology monitoring after the injection of CO_2 could be served as the modification caused by the gas injection. (Morozova et al. 2011). Specifically, the biocenosis such as the sulfate reducing bacteria (SRB) in the rock substrate and fluid samples can be analyzed by the Polymerase chain reaction-Single strand conformation polymorphism method, the Molecular biological method, and the Fluorescent in situ hybridization method (Schilling et al. 2009). It is worth to mention that the information related to the microbiology is valuable for the identification of the biogeochemical process, which influences the diffusion of CO_2 in the reservoirs.

Core analysis is an efficient method for the acquisition of the mechanical and petrophysical properties of formation rock. The measuring methods usually been used for core analysis are the SEM imaging, XRD, and X-ray elemental analysis, which can provide the information of the micro morphological and mineralogical properties of the core (Ringrose et al. 2013).

The satellite-borne Synthetic Aperture Radar (SAR) monitoring can detect and measure the change of ground surface displacements caused by the injection of CO_2, which is beneficial for the modification of the model of underground CO_2 distribution. The mechanism of the SAR is as follows. Firstly, the phase and the amplitude would be obtained by using the SAR. Secondly, the phase difference between two observations will be converted into the ground surface displacements through the look angle and platform altitude (Onuma et al. 2011). Compared the geophysical surveys methods, the technology of satellite-borne SAR monitoring is considered as a cost-effective tool. It is worth to mention that this monitoring technology has been successfully used in the In Salah project (Onuma et al. 2011).

The conventional temperature monitoring cannot provide the high vertical spatial resolution and real-time data. To address these problems, the distributed temperature sensing (DTS) technology is developed. It worth to mention that the fiber optic cable is an important distributed sensor in DTS, which is helpful for the measuring of the temperature from surface to bottomhole along the extension of the fiber (Mawalkar et al. 2019; Shatarah and Olbrycht 2017).

The merits of the monitoring technologies are summarized in Tab. 1.3, and the applications of them in the CCS demonstration projects across the world are listed in Tab. 1.4. The technology of geochemical sampling analyses, 3D seismic, micro-seismic, pressure and temperature logs gain the most popularity among these monitoring technologies. This is result from their excellent performance in acquiring the characteristics of formation fluids and geological structures.

Table 1.3 Main monitoring technologies in CCS (Cao et al. 2020)

Monitoring technology	Advantages	Ref.
3D seismic	Provides tri-dimension image of geological structures and the plume migration of CO_2.	Ringrose et al. 2013
4D seismic	Significant benefits for overburden imaging and time-lapse responses with improved acquisition plan.	Ringrose et al. 2013
Micro-seismic	It is very useful for monitoring the geomechanical response to injection.	Oye et al. 2013
Vertical seismic profiling	Valuable information on the geological structure details.	Götz et al. 2018
Gravimetry	Beneficial for the evaluation of formation fluids density and CO_2 plume.	Kabirzadeh et al. 2017
Cross-hole electromagnetic	Advantageous for the detection and monitoring of the location of CO_2.	Carcione et al. 2012
Pressure and temperature monitoring	Direct information for the evaluation of the stability of the reservoir.	Liebscher et al. 2013
Geochemical sampling	Natural variations in water chemistry are crucial for establishing a useful baseline for groundwater hydrology.	Boreham et al. 2011
Soil and gas sampling	More data on natural CO_2 variations in different environments and associated seasonal fluctuations.	Ringrose et al. 2013
Tracers	Valuable and cost-effective method for monitoring the origin of CO_2 observations at wells and in the storage complex.	Ringrose et al. 2013
Atmospheric monitoring	Useful data to identity the anomalies above the natural baseline.	Etheridge et al. 2011
Microbiology	Valuable data to identify biogeochemical process that affect the diffusion of CO_2 in the reservoirs.	Morozova et al. 2011
Core analysis	Good petrophysical data and rock mechanical properties are essential.	Ringrose et al. 2013
Satellite monitoring	Valuable and cost-effective monitoring data for onshore CO_2 injection operation.	Onuma et al. 2011
Distributed temperature sensing technology	It can provide high-resolution information on the migration of CO_2 in the reservoir.	Mawalkar et al. 2019

Table 1.4 Application of the main monitoring technologies in some CCS demonstration projects (Cao et al. 2020; Leung et al. 2014)

Monitoring technology	Sleipner	Frio	Nagaoka	Ketzin	In-Salah	Otway	Weyburn	MRCSP
3D seismic	×		×	×	×	×		
4D seismic				×	×			
Micro-seismic	×		×		×		×	×
Vertical seismic profiling		×						×
Gravimetry	×				×		×	×
Cross-hole electromagnetic		×		×	×			
Pressure and temperature logs		×	×	×	×			×
Geochemical sampling		×	×		×	×	×	×
Soil and gas sampling		×			×		×	
Tracers		×			×	×		
Atmospheric monitoring						×		
Microbiology				×				
Core analysis					×		×	
Satellite monitoring					×			×
Distributed temperature sensing technology				×				×

1.4.3 Engineering methods to reduce the risks of underground CO_2 storage

► Generating CO_2-in-water foams

Compared with oil and brine, the injected CO_2 exhibits much lower viscosity and density under the reservoir conditions, resulting a poor displacement efficiency. To address this issue, the strategy of generating high viscosity CO_2-in-water foams with large gas volume fraction was proposed (Worthen et al. 2014), which is beneficial for CO_2 storage in both saline aquifers and oil reservoirs. For oil reservoirs, the technology of CO_2-in-water foams can enhance the oil recovery so that improve the economics of CCUS in petroleum systems (Rognmo et al. 2018). Guo and Aryana used a glass microfluidic device to analyze the flow behavior of foam in oil saturated heterogeneous porous medium, and concluded that the foam injection can improve the oil recovery through improving the sweep efficiency (Guo and Aryana 2018). A field experiment of CO_2-in-water foams flooding was conducted in the North Ward-Estes field in Texas. The results demonstrated that the foams can notably improve the sweep efficiency. In addition, this technology can be economically successful (Chou et al. 1992). It should be pointed out that CO_2-in-water foams can also improve the sweep efficiency and storage capacity, as well as economics for saline aquifers. Guo et al. conducted an experimental simulation by using a glass fabricated microfluidic device to investigate the effect of various factors on the CO_2 storage capacity in aquifers (Guo et al. 2019). Their results showed that the CO_2 storage capacity can be increased by over 30% with the help of foam injection compared with the case of CO_2 injection, demonstrating the superiority of CO_2 foam on the improvement of CO_2 storage capacity. It is worth to mention that the CO_2 foam can also reduce the risk of leakage in the underground CO_2 storage unit because of the reduction of fluid mobility. For instance, due to the significant shear rates differences between the flowing in the reservoir formation and leakage pathway, the CO_2 foam may become gel inside the leak. The mechanism is that the particles would break the interaction barrier of forming clusters in high shear stress condition. Therefore, the CO_2 foam can reduce the leakage through the shear-induced gelation. (Pizzocolo et al. 2017).

The surfactants are usually been used to achieve stable CO_2 foams. The hydrophilic/CO_2-philic balance (HCB) is regarded as the principle for the designing of surfactants for CO_2 foams. The HCB can characterize the balance of surfactant and solvent interactions (Johnston and Rocha 2009). In recent years, the nanoparticles combining with the surfactant solutions have been used to improve the stability of CO_2 foams. Meanwhile, the concept of HCB has been developed to be nanoparticle HCB by Worthen et al. (2013). Worthen et al. (2013) generated viscous and stable CO_2-in water foams with the mixture of surfactant (caprylamidopropyl betaine) and nanoparticles (bare colloidal silica). They concluded that the foams were generated by the reduction of interfacial tension through the surfactant. Further, the stability of foams may be enhanced by the adsorption of nanoparticles at the CO_2-water interface. The behavior of silica nanoparticle on the reduction of carbon footprint was also investigated by Rognmo et al. (Rognmo et al. 2018). In addition, various of nanoparticles such as nano Lauramidopropyl Betaine with alpha-Olefin Sulfonate (Guo et al. 2019) also have been used to maintain a high foam quality, demonstrating the superiority of nanoparticles on the stabilizing of CO_2 foams. Overall, the generating of CO_2-in-water foams especially combining with the nanoparticles are considered to be a candidate strategy in the designing of CO_2 storage. It should be mentioned that the overall cost of generating of CO_2-in-water foams should be considered.

► **Accelerating CO_2 dissolution process**

As mentioned in above sections, the free CO_2 would remain more than 1,000 years underneath the caprock, which may lead to some problems. Firstly, it may increase the uncertainties in the long-term fate of injected CO_2. Secondly, it may increase the cost of long-term monitoring operation. To address these issues, the strategy of accelerating the dissolution process of CO_2 was proposed to minimize the free CO_2 in the underground (Anchliya et al. 2012). Cameron and Durlofsky (2012) used the Hooke–Jeeves Direct Search algorithm to optimize the injection rate and the locations of CO_2 injection wells to minimize the mobile CO_2 in the CCS units. Their results showed that the fraction of mobile CO_2 decreases from 0.220 to 0.072 in the optimal case, demonstrating the importance of well location optimization. Anchliya et al. (2012) developed an engineered injection method to accelerate the dissolution and trapping of CO_2. The schematic of the engineered injection process was shown in Fig. 1.8. Compared with conventional CO_2 injection scenarios, there exits another brine injection well that is located near the top of reservoir and exactly over the horizontal CO_2 injection well. In this engineered injection system, the other two brine production wells are placed at either side of the CO_2 injection well. It should be pointed out that the brine injection well is used to limit the upward movement. In addition, the brine injection well can impel the horizontal flows of CO_2 under a lateral pressure gradient provided by brine injection. Therefore, the sweep efficiency would be increased and the CO_2 dissolution and trapping would be enhanced (Anchliya et al. 2012). Numerical simulation studies show that around 90% of the injected CO_2 can be immobilized within 20 years after the ceases of CO_2 injection because of the fast dissolution trapping and residual trapping. By controlling the brine injection and production rates, the potential risk of pressurization caused by CO_2 injection could also be addressed by this engineered injection method. Further, it can increase the storage capacity of CO_2 in a bounded aquifer formation. However, it should be mentioned that additional drilled wells are needed for the engineered injection system, which increases the cost of this CCS operation. Consequently, the adaptability of this engineered injection method should be analyzed and quantified site specific.

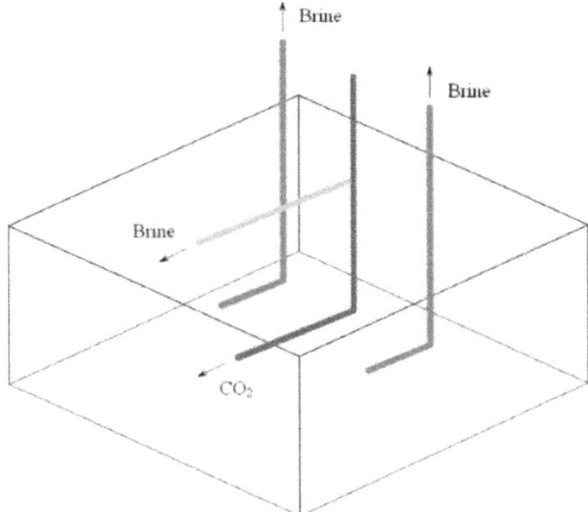

Figure 1.8 Engineered injection method to accelerate CO_2 dissolution and trapping (Anchliya et al. 2012)

Another injection scheme called water-alternating-gas (WAG) injection was firstly developed in the petroleum industry in the late 1950's to improve the sweep efficiency of reservoirs. Seo et al. (2019) performed experimental studies to investigate the performance of sequential water injection associated with gaseous CO_2 into brine, and concluded that this injection scheme is beneficial for minimizing the drying-out of brine and the precipitating of salts at pores, as well as accelerating the dissolution of CO_2. The different schemes of WAG injection are illustrated in Fig. 1.9. The performance of the WAG injection scheme with a goal of improving the efficiency of CO_2 storage was investigated by Zhang and Agarwal (2012, 2013). Their results showed that the optimized WAG scheme can accelerate CO_2 dissolution and decrease the impact zone up to 14% compared with that of the constant gas injection scheme. However, the WAG injection scheme may decrease the total storage capacity of the reservoirs because of the large amount of injected water. In addition, it may increase the overall cost in the injection process, thus it has not been applied widely in the CCS operation.

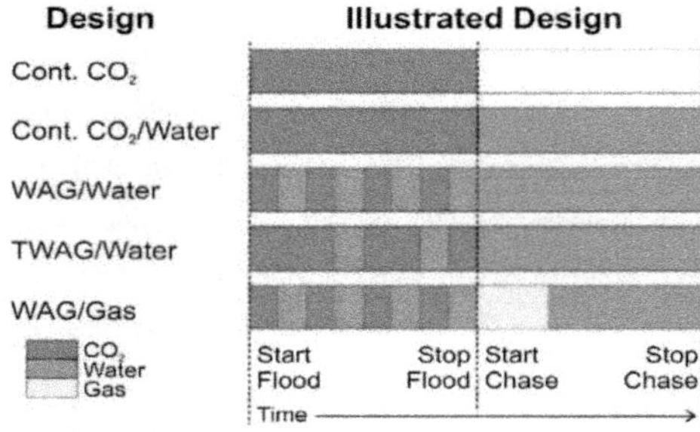

Figure 1.9 Schematic of various schemes of WAG injection (Harris et al. 2009)

To mitigate the adverse impacts of WAG injection schemes on storage capacity, an injection scheme combing the intermittent injection method and brine production was proposed by Tanaka et al. (2013). The schematic diagram of the intermittent injection method is shown in Fig. 1.10, which suggests that a diagonal pair of wells are used for CO_2 injection alternately. In this injection scheme, a diagonal pair of wells (well 1 and well 3) are used for CO_2 injection, while another pair of wells (well 2 and well 4) are used for brine production, which will be re-injected into the reservoir through well 1 and well 3. Numerical simulation results reveal that both the dissolved and residual CO_2 are increased compared with the base case that has only one well with continuous injection. Specifically, the ratio of the trapped CO_2 increases by 20%. It is worth to mention that this intermittent injection scheme can mitigate the pressure buildup through intermittent injection and water production, demonstrating the importance of brine management and the CO_2 injection configuration.

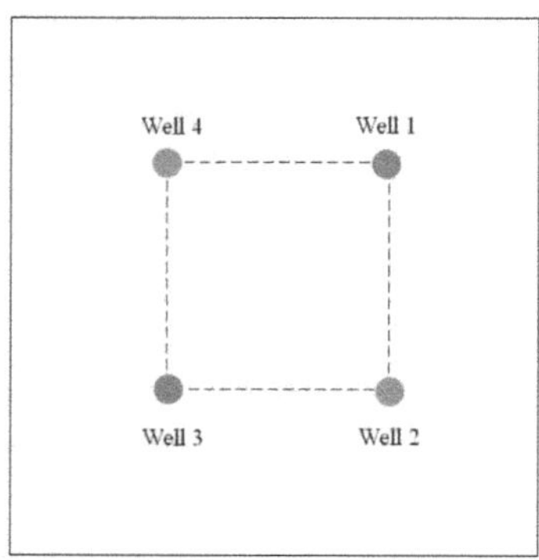

Figure 1.10 Intermittent injection method (Cao et al. 2020; Tanaka et al. 2013)

Regarding above methods for accelerating the dissolution of CO_2, they may increase the cost in the injection process due to water injection. In addition, additional wells are needed for the engineered injection and intermittent injection, resulting high costs on the operation. However, the cost on the monitoring would be reduced account for the relatively low leakage risk caused by the rapid dissolution rate of CO_2. The overall economic costs of the CCS unit and the storage capacity are suggested to be taken into consideration for optimization when considering the utilization of these engineering methods.

► Accelerating mineral carbonation process

The carbonation of CO_2 is an effective way to fix the injected CO_2 and guarantee its security permanently. However, for conventional CCS in saline aquifers with silicate minerals, it takes tens of thousands of years for the mineralization account for the low reactivity of silicate minerals in sedimentary rocks with CO_2 (Metz et al. 2005). To accelerate the mineralization process of CO_2, some novel methods have been proposed, including CO_2 storage in basalt rock formation (Gislason and Oelkers 2014), CO_2 storage in peridotite formation (Kelemen and Matter 2008), direct mineralization of flue gas by coal fly ash (Reddy et al. 2011), direct aqueous mineral carbonation (Verduyn et al. 2011), and pH swing mineralization (Wang and Maroto-Valer 2013).

• CO_2 storage in basalt rock formation

The method of applying CO_2 sequestration in basalt rock formation is proposed by Gislason and Oelkers (2014). Basalt contains around 25% of magnesium, calcium, and iron oxides, which are far more reactive with carbonic water compared with silicate minerals in sedimentary rock (Schaef et al. 2010). Therefore, it may cost less time for the carbonation of injected CO_2 in basalt rock formation. It should be mentioned that that there exist abounding basaltic rocks on the earth's surface (Goldberg et al. 2008), which offers the possibility of CCS in basalt formation in large scale. A field test of CCS in basalt formation was conducted in the CarbFix pilot project in Iceland in 2012 (Matter et al. 2016). In this project, 175 tons of pure CO_2 were injected into the basalt formation with water firstly, and then 73 tons of CO_2-H_2S mixture (55 tons CO_2) were fully dissolved in water and injected. The data of the measured dissolved inorganic carbon of ^{14}C at the monitoring well shows that more than 95% of injected CO_2 was mineralized to carbonate minerals within 2 years, showing the efficiency of mineral trapping of CO_2 in basalt rock. However, it should be pointed out that the large-scale application of this technology requires substantial quantities of water during the CO_2 injection process. In addition, the cost of storage and transportation of CO_2 for the CarbFix project is about twice compared with that of in typical sedimentary basins. Therefore, the application prospect of CO_2 storage in basalt rock formation is not very promising in the near future.

- **CO_2 storage in peridotite formation**

The mantle peridotite is mainly composed of olivine and pyroxene, which can react with CO_2 and H_2O with the formation of hydrous silicate, Fe-oxides, and carbonates. The data of isotope analysis and reconnaissance mapping indicate that approximately 10^4 to 10^5 tons of CO_2 per year are trapped to solid minerals through the peridotite weathering effect in Oman (Kelemen and Matter 2008). The main reaction of CO_2 storage in peridotite formation can be expressed as:

$$2Mg_2SiO_4 + Mg_2Si_2O_6 + 4H_2O = 2Mg_3Si_2O_5(OH)_4 \tag{1.2}$$

$$Mg_2SiO_4 + 2CO_2 = 2MgCO_3 + SiO_2 \tag{1.3}$$

$$Mg_2SiO_4 + CaMgSi_2O_6 + 2CO_2 + 2H_2O = Mg_3Si_2O_5(OH)_4 + CaCO_3 + MgCO_3 \tag{1.4}$$

As shown in Fig. 1.11, the carbonation rate can be enhanced more than 1 million times than natural rate with the help of the 3-step operation process. This operation begins with drilling and fracturing, followed by the injection of hot CO_2 (approx. 185 °C) at a rapid rate to heat the fractured peridotite. The last step is injecting CO_2 with normal temperature. During the chemical reaction in this case, the system would maintain at a temperature of 185 °C and high carbonation rate due to the exothermic carbonation of peridotites. It is estimated that the peridotite in Oman alone could trap more than 1 billion tons of CO_2 per year into carbonate minerals, showing a huge CO_2 storage capacity.

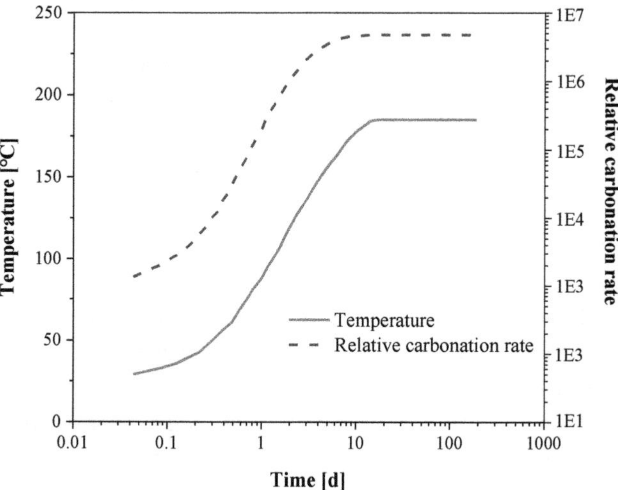

Figure 1.11 Calculated temperature, and the carbonation rate relative to the rate for CO_2 in surface water at 25°C and 0.1 MPa in the 3-step injection operation (adapted from Kelemen and Matter 2008)

However, except for the peridotite exposing through large thrust faults, the mantle peridotite is generally beneath more than 6 and 40 km below the seafloor and land surface respectively (Kelemen and Matter 2008), making it difficult for the application of CO_2 sequestration in these formations. It is worth to pointed out that the peridotite in shallow formations can be used effectively for the sequestration of CO_2.

- **Direct mineralization of flue gas by coal fly ash**

Reddy et al. (2011) conducted a preliminary experiment to investigate the reaction between flue gas and coal fly ash in a fluidized bed reactor. Fig. 1.12 shows the diagrammatic sketch of the preliminary experimental setup. The experimental results show that the concentrations of CO_2 and SO_2 in flue gas decreased from 13.0% to 9.6%, from 107.8 to 15.1 ppmv, respectively within 2 minutes. Furthermore, the Hg in flue gas was also mineralized by the fly ash particles. Reddy et al. (2011) conducted a pilot scale study with a fly ash content of 100-300 kg to analyze the feasibility of this kind of technology. In the pilot studies, the fly ash particles were fluidized by the flowing of flue gas in the fluidized bed reactor to ensure sufficient mixing and contact between them. The reaction occurred under a fixed pressure of 115.1 kPa. According to the experimental results, the content of $CaCO_3$ produced by the reaction of flue gas and fly ash ranged from 2.5% to 4% in 10 minutes. Meanwhile, the contents of S and Hg in the fly ash increased from non-detectable to 0.45 and 0.5 mg/kg, respectively. These results confirmed that the flue gas components can be captured without separation and mineralized by fly ash particles by using the method of accelerating mineral carbonation process. However, it should be pointed out that the treatment of the carbonated fly ash produced by using this method is still a crucial problem that needs to be addressed (Wee 2013).

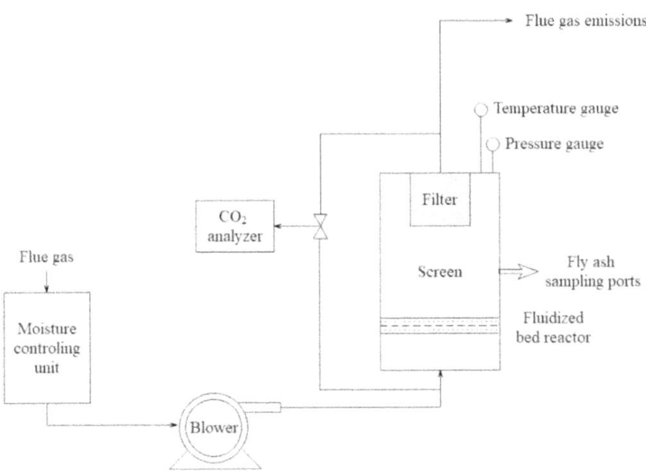

Figure 1.12 Preliminary experimental setup for CO_2 capture and mineralization (Cao et al. 2020; Reddy et al. 2011)

- **Direct aqueous mineral carbonation**

Direct aqueous mineral carbonation is a method that uses a bicarbonate-bearing solution mixed with reactant minerals such as magnesium and calcium silicate rocks, to convert gaseous CO_2 into solid form (O'Connor et al. 2001). It should be mentioned that the magnesium and calcium silicate rocks are distributed all over the world with an abundant reserve. For instance, it is estimated that the magnesium silicate rock in Eastern Finland has the ability to store 10 million tons of CO_2 per year for 200 to 300 years (Teir et al. 2009).

The overall chemical reaction of the carbonation of serpentine can be expressed as:

$$Mg_3Si_2O_5(OH)_4 + 3CO_2 = 3MgCO_3 + 2SiO_2 + 2H_2O \quad (1.5)$$

In the carbonation process, the serpentine was treated at 630 °C to attain an active mineral. Then the stoichiometric conversion of 78% for the silicate to carbonate was observed by using the bicarbonate-bearing solution at the conditions of 150 °C and 18.5 MPa, with 15% solids. It is worth to mention that the conversion occurred within 30 minutes, showing an extremely rapid reaction rate for the carbon mineralization. However, there is massive energy consumption related to this CCS technology because of the requirement of heat treating. To reduce the energy consumption on the heat reactivation and increase the effective reaction area, the mechanical activation was proposed by adding an attrition grinding step. In this case, the reaction condition was optimized to 25 °C and 1 MPa. According to the result, it still led to up to 65% carbonation within 1 hour (O'Connor et al. 2001), also showing a high reaction rate. Further, the combination of attrition grinding and heat activation was used to achieve a better carbonation mineralization performance, as shown in Fig. 1.13a (Verduyn et al. 2011). Based on this concept, the feasibility of this technology in the mineralization of flue gas was analyzed by Verduyn et al. (2011). As shown in Fig. 1.13b, the relatively high pH due to the low solubility of flue gas makes it difficult for the leaching of cations. To eliminate this problem, the contact of the mineral and flue gas is done in a slurry mill and a leaching basin.

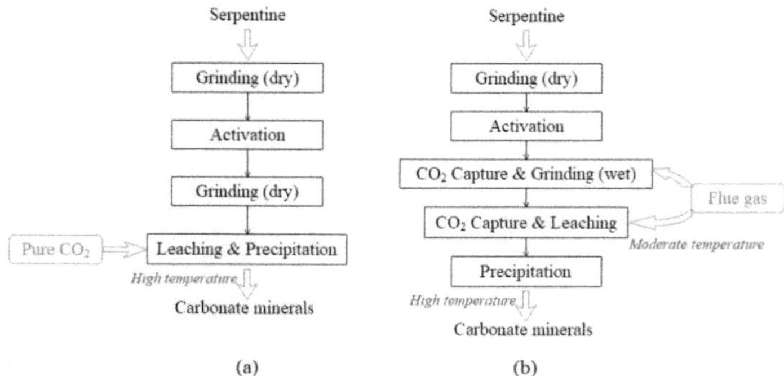

Figure 1.13 Mineralization concept for pure CO_2 and flue gas (Cao et al. 2020; Verduyn et al. 2011)

- **pH swing mineralization**

To increase the conversion efficiency of CO_2 mineralization, the pH swing approach was proposed by Park and Fan (2004). They combined the internal grinding in acidic solvent for a rapid dissolution of a serpentine sample. According to the results, three solid products including SiO_2-rich solids, iron oxide, and magnesium carbonate were produced by controlling the pH. Teir et al. (2009) used HCl and HNO_3 to dissolve serpentinite, and then transform the serpentinite to hydromagnesite with the help of CO_2. In their studies, the pure hydromagnesite which is thermally stable at 300 °C was produced. However, the additional amount of chemicals used in the operation increase the overall costs and make it infeasible for the application. To reduce the cost of pH swing CO_2 mineralization, a recyclable reaction solution was introduced by Kodama et al. (2008). They selectively extracted the alkaline-earth metal from steelmaking slag in an ammonium chloride solution. The reacted solution was used for CO_2 absorbent and ammonium carbonate production. Further, the calcium carbonate was precipitated in another reactor with the recovery of ammonium chloride. The results revealed that the selectivity of the calcium extraction reaction reached 60%. In addition, pure $CaCO_3$ was produced with an energy consumption of 300 kWh/t-CO_2. Wang and Maroto-Valer (2013) developed a modified carbon mineralization process, the schematic diagram was shown in Fig. 1.14. Through experimental studies, Wang and Maroto-Valer (2013) concluded that $(NH_4)_2CO_3$ is more favorable for increasing the efficiency of carbon fixation compared with NH_4HCO_3, and the optimal efficiency of CO_2 mineralization reaches to 46.6%. Further, the pH swing mineralization process was optimized by Sanna et al. (2013) under different temperature conditions. The results showed that the total CO_2 trapping efficiency was 62.6% at the temperature of 80 °C, with the molar ratio of 1:4:3 for Mg: NH_4 salts: NH_3. However, it should be pointed out that the energy consumption and overall economic cost are supposed to be lowered before any large-scale application.

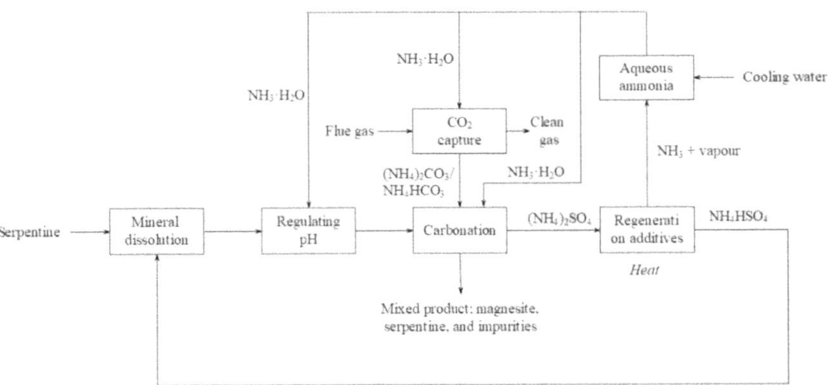

Figure 1.14 The schematic of carbon mineralization process using recyclable ammonium salts (Cao et al. 2020; Wang and Maroto-Valer 2013)

In summary of this section, the implementation of CO_2 sequestration in basalt rock formation and peridotite formation is limited by the distribution of the particular rock. In addition, the substantial quantities of water required for dissolution and the energy consumption required for heating increase the overall cost a lot. Due to the fact that the flue gas can be mineralized without separation, which decreases the overall cost a lot, it can be inferred that the direct mineralization of flue gas by coal fly ash would be a promising technology. The pH swing mineralization may be another promising technology for the mineral trapping of CO_2 because of its sustainability. It is suggested to introduce recyclable and cheap chemical reagents into the mineralization process to make this technology more cost-effectiveness.

1.4.4 Uncertainty analysis for CO_2 storage

There are numerous origins of uncertainty during the process of CCS operation, such as the measuring uncertainty of critical mechanical and geological parameters, and the scaling uncertainty of upscaling from a pilot project to an industrial scale CCS project. These uncertainties would probably reduce the accuracy of feasibility assessments, as well as the evaluation of CO_2 storage capacity in CCS units (Chadwick et al. 2008; Jayne et al. 2018; Pan et al. 2016). Uncertainty analysis can mitigate the harmful impact and assist in avoiding potential accidents and economic losses. Therefore, uncertainty analysis is very important for the systematic assessment of CCS units.

The effect of uncertainty parameters on CCS operation has been investigated by researchers in recent years. Sarkarfarshi et al. (2014) analyzed the effect of different parameters with specific uncertainties on CO_2 plume evolution. Their results showed that formation porosity and brine saturation are two critical parameters in this process, because they have a significant impact on the absolute permeability of reservoir and the relative permeability of CO_2, respectively. Li and Laloui (2017) developed a thermal-hydrological-mechanical (THM) numerical model to investigate the impact of coupled material properties on caprock stability. The material properties considered in the study including the Young's modulus, Biot's coefficient, and thermal expansion coefficient. The numerical results indicated that the risks of the failure in caprock may increase or decrease at a temperature difference of 30 °C between the injected CO_2 and the reservoir. They inferred that this contradictory behavior is caused by the coupling effect of the thermal-hydromechanical parameters. Therefore, the coupling effect of the material properties on the caprock are supposed to be taken into consideration, especially when CO_2 is injected at a low temperature. The geomechanical response of CCS unit for a fluctuating injection rate was investigated by Bao et al. (2014). According to their results, the maximum sustainable injection pressure decreases when fluctuation exits in the injection rate. Hou et al. (2014) performed an uncertainty analysis on CO_2 plume expansion subsequent to wellbore leakage. Newell et al. (2017) conducted a comparative study on the effect of different geomechanical and hydrogeological properties on the surface uplift and pore pressure based on the In Salah project. The simulation results showed that a better match with the surface displacement can be attained if Biot's coefficient was taken into account, demonstrating the importance of Biot's coefficient under a hydraulic-mechanical coupled framework in CCS units. Jayne et al. (2019) investigated the effect of permeability uncertainty in a basalt-hosted CCS reservoir with the help of a stochastically generated and spatially correlated permeability distribution. Their results revealed that the ensemble variance exhibits an ellipse uncertainty around the CO_2 plume. This indicated that the uncertainty of the permeability affects both the distribution and accumulation of CO_2 in significant. These studies laid the foundations for the analysis on the role of formation parameters on CCS. However, the uncertainties related to the geomechanical and hydrogeological parameters have not been fully considered, even though they have a visible impact on the response of formation in CCS operation. In addition, the results obtained from the site-specific uncertainties of these parameters cannot be identified as a general law for common CCS operation.

In recent years, many mathematical and computational methods have been introduced into the evaluation and quantification of the uncertainties associated with CO_2 sequestration. Bao et al. (2013) quantified the impact of the reservoir and caprock property uncertainties on pressure buildup and surface uplift during CCS operation with the help of Quasi-Monte Carlo method, generalized cross-validation and analysis of variance method, in which the former is an efficient approach for the sampling of the parameters with high dimensions. A probability density function was introduced by Wriedt et al. (2014) to analyze the potential risks during the process of CO_2 sequestration. It is worth to mention that the probability density function can also be used for other reservoirs with uncertainty parameters. The response surface method is a classical statistical method, which can be used for optimization based on the objective function. By using this method, according to the basic data in the Farnsworth Unit enhanced oil recovery field in Texas, Pan et al. (2016) concluded that the maximum cumulative oil production can be attained with a permeability in the range of 10–31.6 mD under the condition of constant CO_2 injection. Namhata et al. (2016) developed an arbitrary polynomial chaos expansion method. Specifically, a massive stochastic model reduction was used to investigate the uncertainties in pressure prediction caused by leakage at the zone overlaying caprock. For the purpose of minimizing the computational cost, Jeong and Srinivasan (2016) developed a fast-alternative method that can be used to quickly quantify the uncertainty of both the spatial and temporal features of CO_2 plume migration. Shirangi and Durlofsky (2016) generated a framework for representative subset selection in a large set. The selection was performed by using principal component analysis and a clustering algorithm, which is beneficial for decision making with uncertainty. Due to the superior performance in terms of sensitivity analysis of the multivariate adaptive regression spline method, it was used by Dai et al. (2018) to perforem a global sensitivity analysis on CO_2 storage in marine sediments. More recently, Singh (2019) introduced machine learning technology to investigate the surveillance of fluid leakage from the reservoir in relation to bottom-hole pressure and injection rate. These studies offer valuable insights into the uncertainty quantification of CCS units. However, to the best of our knowledge, there are few studies focus on machine learning approaches for the investigation of the patterns of formation response including formation fluid pressure and formation displacement caused by CO_2 injection.

1.5 Methods for improving the cost-effectiveness of CO_2 storage

1.5.1 Enhanced industrial production with CO_2 storage

Resources production during CO_2 storage is considered as an effective method to partly cover the cost of CCS, which is so called CO_2 capture, utilization, and storage (CCUS) (Liu et al. 2017). It should be pointed out that it may potentially achieve additional economic benefits. During the process of CCUS, CO_2 usually works as working fluid to enhance the recovery of underground resources. The mechanisms of enhance the recovery of underground resources including dissolution, displacement, reactive transport, and thermal conductivity. The potential geological formations for CCUS include oil reservoirs, gas reservoirs, saline aquifers, shale formation, un-mineable coal seams, hot dry rock, uranium deposit formation, and natural gas hydrate reservoirs (Wei et al. 2015). The corresponding CCUS technologies are CO_2 enhanced oil recovery (CO_2-EOR), CO_2 enhanced gas recovery (CO_2-EGR), CO_2 enhanced water recovery (CO_2-EWR), CO_2 enhanced shale gas recovery (CO_2-ESGR), CO_2 enhanced coalbed methane recovery (CO_2-ECBM), CO_2 enhanced geothermal systems (CO_2-EGS), CO_2 enhanced in situ uranium leaching (CO_2-IUL), and CH_4-CO_2 replacement from natural gas hydrates, respectively (Burton et al. 2013; Wei et al. 2015).

► CO_2-EOR

The technology of CO_2-EOR is the most successful and promising technology combining the utilization and storage of CO_2, in which CO_2 is injected into oil reservoirs to enhance the recovery of crude oil (Azzolina et al. 2016; Bian et al. 2016; Singh and Haines 2014). The displacement mechanisms of CO_2-EOR can be classified as multicontact miscible and immiscible processes, depending on the properties of the reservoir fluids and CO_2 under the condition of reservoir pressure and temperature (Brush et al. 2000; Metcalfe 1982). Regarding the multicontact miscible displacement, the minimum miscibility pressure (MMP) is required. When the pressure is lower than the MMP, the immiscible displacement occurs with less components exchange between oil and CO_2 in the reservoir (Sahin et al. 2007).

There are three CO_2 injection methods for the operation of CO_2-EOR, including continuous injection, water alternating gas (WAG) injection, and cyclic injection. Regarding the continuous injection method, CO_2 injection and oil production are running continuously. This kind of CO_2 injection method has been applied in the North Cross Devonian Unit for enhanced oil recovery (Aryana and Barclay 2014). The multicontact process can be achieved through vaporizing and condensing (Jia et al. 2019). However, this injection method has not gained much popularity in the field application compared with the WAG injection and cyclic injection. WAG injection has been widely used because it can decrease the mobility ratio between the injection fluids with oil and lead to late gas breakthrough and high oil recovery. In the design of WAG injection, the optimization algorithm such as the Lagrangian and stochastic simplex approximate gradient algorithm can be used to achieve the maximum net present value (Chen and Reynolds 2017). Although the WAG injection is considered as an effective method to improve the oil recovery, it may cause the gas flowing upward while the water and oil flowing downward because of the large density differences. This may lead to early gas breakthrough, especially in the reservoir with large vertical heterogeneity and highly permeable channels (Jia et al. 2019). To address this problem, the cyclic injection process, i.e., gas huff-n-puff process was proposed. The gas huff-n-puff is composed of three stages, including the gas injection stage, well shutting state, and the oil production stage.

For conventional oil reservoirs, CO_2 flowing is dominated by the rock matrix. The mechanism of CO_2-EOR is due to the solubility of CO_2 in oil under the supercritical phase condition, which can decrease the density and viscosity of oil, increasing the mobility of oil and resulting enhanced oil recovery (Jia et al. 2019). For the unconventional tight oil reservoirs such as shale oil reservoirs, fracturing is an essential technology for the exploitation. In this scenario, CO_2 flowing is dominated by fracture flow instead of rock matrix flow. As can be seen in Fig. 1.15, the process of CO_2-EOR in fractured tight oil reservoirs can be divided into 4 steps. In the first stage (step 1), the injected CO_2 flows rapidly through the fracture. Then the CO_2 starts to permeate into the rock matrix under the displacement effect (step 2). During this stage, the permeating CO_2 may carry oil into the rock and decrease the oil production. Simultaneously, the permeated CO_2 would lead to the swelling of oil and then mitigating out of the matrix, which is in favor of the oil production. The oil continues to swell with the decreasing of viscosity caused by the permeated CO_2. Further, the oil would move to the fracture in the follow stage (step 3), which corresponding to the well shutting stage in the huff-n-puff process. Finally, the pressure equilibrium inside of the matrix would be achieved, thus the migrating of the miscible or immiscible oil from the matrix to the fracture is dominated by the diffusion effect. The oil in the bulk CO_2 is migrating along the fractures to the production well under the effect of pressure gradient (Hawthorne et al. 2013). It should be mentioned that the cyclic injection scheme can also promote the propagating of reservoir pressure because of CO_2 injection near the injection well, especially for the reservoir with ultra-low permeability. For instance, the CO_2 huff-n-puff shows better performance on the oil recovery when the reservoir permeability is lower than 0.03 mD (Yu et al. 2017). In the future studies, characterization of the flow behavior of CO_2 and oil with accuracy in the low permeability reservoir with complex natural and hydraulically created fractures under the in-situ temperature and pressure conditions are encouraged to be emphasized, which is in favor of improving the efficiency of CO_2-EOR.

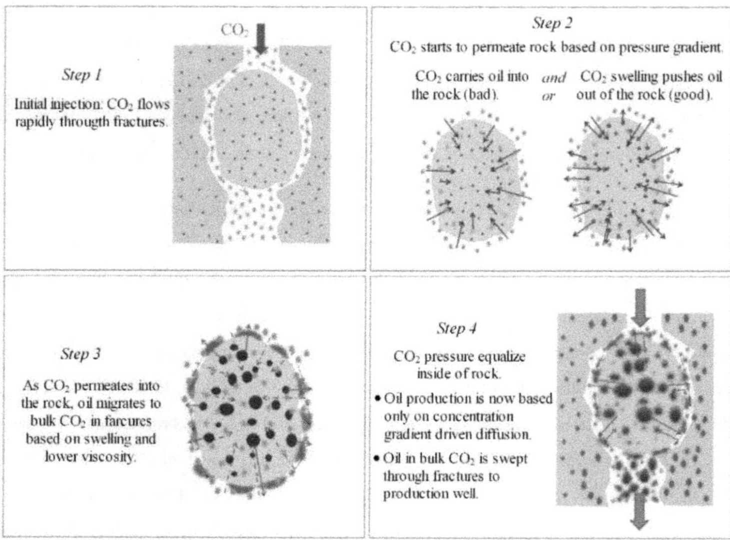

Figure 1.15 Conceptual steps of CO_2-EOR in fractured tight oil reservoirs (Hawthorne et al. 2013)

Apart from increasing oil recovery, CO_2-EOR can also play a role on CO_2 sequestration. It is supposed that CO_2-EOR could be an important economic incentive for early CO_2 storage projects (Mac Dowell et al. 2017). Typically, 3 tons of CO_2 injection can produce about 1 bbl of incremental oil. It is reported that approximately 5% to 15% enhancement in oil production can be obtained through the technology of CO_2-EOR (Bui et al. 2018). In the largest discovered oil fields all over the world, it is estimated that approximately 470 billion barrels of incremental oil can be produced simultaneously with 140 billion metric tons of CO_2 stored by using CO_2-EOR (Carpenter and Koperna 2014).

The first CO_2-EOR pilot project was implemented at the SACROC oil field in 1972 (Kane 1979). In this project, CO_2 foam was used to alter the mobility of oil and improve the sweep efficiency (Langston et al. 1988; Sanders et al. 2012). At present, the technology of CO_2-EOR is relatively mature and has been widely used in petroleum industry to enhance oil recovery for tens of years, with the capacity of more than 1,000 million tons of CO_2 stored subsurface (Gozalpour et al. 2006; Marston 2017). Especially, CO_2-EOR has gained great success in North America. In the USA, the oil with a rate of more than 260,000 bbl/d are produced account for the application of CO_2-EOR technology (Clemens et al. 2010). In the Weyburn oilfield in Canada, the CO_2-EOR project was implemented to extend the life of the oilfield. A total of approximately 20 million tons of CO_2 is scheduled to be stored in the oil reservoir (Hattenbach et al. 1998; Preston et al. 2005). In recent years, the feasibility of CO_2-EOR in China has been massively studied. The first CO_2-EOR project in China, i.e., Jilin Oilfield, has been injected nearly 217,000 tons of CO_2 with a storage efficiency of 96% by April, 2013 (Lui and Leamon 2014), and the total CO_2 storage capacity is designed as about 600,000 tons (Global CCS Institute 2019). It is reported that the technology of CO_2-EOR has application prospects in the Shengli Oilfield and Bohai Bay Basin, which can obtain 6.7% incremental oil recovery and 683 million tons of incremental oil production, respectively (Lv et al. 2015; Yang et al. 2017). The technology of CO_2-EOR also attracted much attention in Europe. For example, the potential utilization of anthropogenic CO_2 for CO_2-EOR was studied based on the B8 oilfields located at the Baltic Sea and Brage on the Norwegian Continental Shelf (Mathisen and Skagestad 2017), which is a part of the ongoing PRO_CCS project funded by Norway Grants. The simulation results indicated that the total CO_2 storage capacity of Brage and B8 oilfields are 33 and 4.8 million tons respectively in 17 years of injection. Simultaneously, an expected incremental oil production of 98 and 14.6 million bbls could be achieved in the two oilfields, respectively.

For the purpose of optimizing the CO_2 storage and enhanced oil recovery in CO_2-EOR operation, Ampomah et al. (2016) proposed an objective function (Eq. 1.6) with consideration of both CO_2 storage and oil production. It should be mentioned that the objective function can be optimized by the neural network and genetic algorithm. In the case study of the Farnsworth field unit, more than 94% of CO_2 can be stored with about 80% of oil produced. This lays a foundation for the co-optimization of CO_2 storage and EOR.

$$f = w_1 \times FOPT + w_2 \times FGIT \tag{1.6}$$

Where w is weight assigned to vector, *FOPT* is the cumulative produced gas, and *FGIT* is the cumulative injected gas.

Similarly, a framework with the objective of co-optimize CO_2 storage and oil production was developed by Jahangiri and Zhang (2012). In the framework, the net present value (NPV) was treated as the optimization objection function, which can be solved by the ensemble-based optimization algorithm as written in Eq. 1.7.

$$\text{NPV} = \sum_{t=1}^{T} \frac{C}{(1+r)^t} - C_0 \tag{1.7}$$

Where t is the time step, T is the operation period, r is the periodic discount rate, C is the cash flow in the time step that is determined by price, injection and production volume of CO_2 and oil, C_0 is the initial investment.

By using the above method, the injection rates and well injection patterns corresponding the maximum NPV can be determined. In addition, the discrete time optimization model can be applied for maximizing the total profit in CO_2-EOR operations, with both enhanced oil recovery and geological CO_2 sequestration considered (Tapia et al. 2014).

It should be pointed out that the artificial neural network models can be used to predict and optimize the performances of CO_2-EOR. Fig. 1.16 shows the artificial neural network structure of the models during the multi-cycled water-alternating-gas process (Van and Chon 2017). There are four neurons in the input layer corresponding to permeability ratio, initial water saturation, water-to-gas injection time ratios (WAG), and temperature, respectively. Subsequently, the oil production rate, oil recovery, net CO_2 storage amounts, and gas and oil ratio (GOR) are set as the targets, which are corresponding to the four neurons in the output layer. It should be mentioned that there are 10 neurons in the hidden layer. The oil recovery and net CO_2 storage can be accurately predicted based on this framework. For instance, the optimal injection scheme for the maximum economic profit can be obtained in various reservoir conditions by using this method.

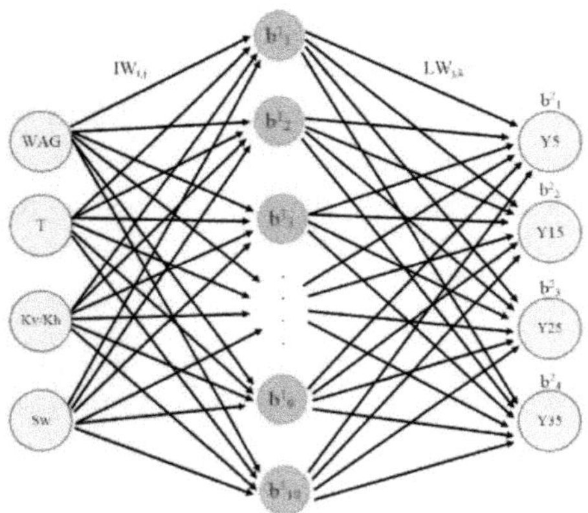

Figure 1.16 Artificial neural network structure of the models, Y represents oil recovery, oil production rate, GOR and net CO_2 storage amount (Van and Chon 2017)

The machine learning approach can also be used to optimize the oil recovery and stored CO_2 in the CO_2-EOR operation (You et al. 2019). Under this optimization framework, a history matching model was developed based on the production history data in the CO_2-EOR process. The flowchart of the optimization framework is shown in Fig. 1.17. The hybridized multi-layer and radial basis function Neural Network method were utilized to train a proxy model, which is beneficial for increasing the computational effectiveness during the optimization process. After a proxy model with reliable accuracy was obtained, the machine learning optimization algorithm was used to acquire the optimal solution of the objective function that incorporates the role of parameters such as the oil recovery and stored CO_2. This research work demonstrates the adaptability of robust machine learning approach on optimizing the CO_2-EOR process. Generally, considering the maturity of the technology and huge market demand, it is estimated that CO_2-EOR may play a more important role on mitigating CO_2 emissions than other strategies of CO_2 utilization in the next few years.

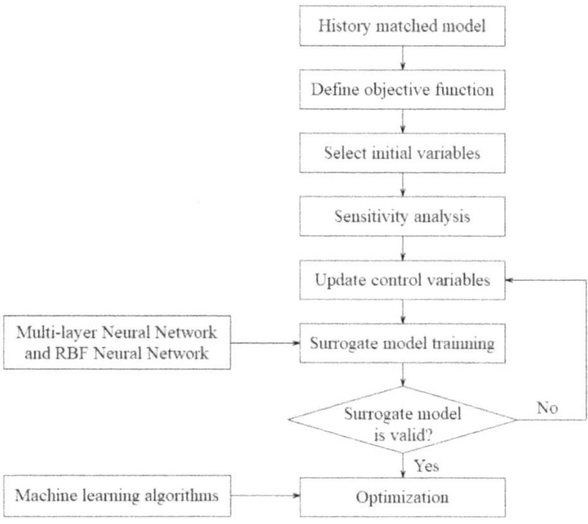

Figure 1.17 Flowchart of the optimization framework (Cao et al. 2020; You et al. 2019)

► CO_2-EGR

CO_2-EGR means that it enhances the gas recovery through the injection of CO_2. The gas recovery is enhanced by both displacement effect and re-pressurization of the remaining gas in a depleting or depleted reservoir (Al-Hasami et al. 2005). Regarding the sour gas reservoirs that CO_2 is produced mixed with the natural gas. the separated CO_2 from the produced gas can be injected back into the reservoir to enhance gas recovery. Additionally, it should be mentioned that CO_2 has the potential to decrease the dew point pressure of reservoir fluids in wet gas reservoirs, which is favorable for eliminating condensate blockage and simultaneously improving CH_4 production (Odi 2012, 2013). It is estimated that up to 11% incremental gas recovery can be achieved by CO_2 injection (Al-Hasami et al. 2005).

The feasibility of CO_2-EGR has been analyzed by many experimental and numerical simulation studies (Clemens et al. 2010; Eliebid et al. 2018; Khan et al. 2012, 2013; Klimkowski et al. 2015; Mamora and Seo 2002; Narinesingh and Alexander 2014; Odi 2012; Seo and Mamora 2005; Zangeneh and Safarzadeh 2017). Some typical displacement experiments in a variety of temperature and pressure conditions are summarized in Tab. 1.5. Abdoulghafour et al. (2016) pointed out that the residual water narrower the pore and consequently increases the dispersion of supercritical CO_2 and CH_4. Abba et al. (2017) resulted that the salinity of connate water will decrease the dispersion of CO_2 in CH_4. Abba et al. (2018) concluded that the gravity has significant effects on the flow behavior of SCO_2 at lower flowrates. These observations reveal the mechanism of CO_2-EGR and provide a guide line for the application of this kind of technology.

Table 1.5 Typical displacement experiments on CO_2-EGR process (Cao et al. 2020)

Rock type	Saturated fluids	T (°C)	P (MPa)	Key observations	Reference
Carbonate core	CH_4	20~60	3.55~20.79	Whether CO_2 is gas, liquid, or supercritical phase, it could enhance the recovery of CH_4.	Mamora and Seo 2002
Carbonate core	saturated with methane with or without water	20~80	3.55~20.79	The coefficient of CO_2 increases with temperature and decreases with pressure.	Seo 2004
Berea Sandstone core	dry core, initial saturation of 10% distilled water, and initial saturation of 10% brine (20 wt%), respectively	40	8.96	The salinity of connate water will decrease the dispersion of CO_2 in CH_4.	Abba et al. 2017
Sandstone & carbonate core	CH_4	60~80	10~12	The residual water narrower the pore and consequently increases the dispersion of supercritical CO_2 and CH_4.	Abdoulghafour et al. 2016
Sandstone core	CH_4 and simulate natural gas (90% CH_4 + 10% CO_2) respectively	40~55	10~14	The dispersion coefficient of CO_2 in the simulate natural gas is larger than that of CH_4.	Liu et al. 2018
Sandstone core	formation water and N_2	50	21	The gravity segregation effect is notable in the porous and permeable core, while heterogeneity effect becomes dominant in the low permeability core.	Liu et al. 2015
Bandera sandstone core	CH_4	50	8.96	The gravity has significant effects on the flow behavior of SCO_2 at lower flowrates.	Abba et al. 2018

The most critical problem in CO_2-EGR is the breakthrough of CO_2 in the reservoir, resulting the production of CO_2 contaminated gas (Khan et al. 2012; Zangeneh et al. 2013). Actually, the preferential pathway has significant effect on CO_2 breakthrough and ultimate CH_4 recovery (Wang et al. 2013). Therefore, the geological formations, especially the microstructures, are supposed to be characterized in detail for accurately predicting the behavior of CO_2 flowing. The irreducible water in reservoirs also affect the mixing of CO_2 and CH_4 (Honari et al. 2016). The dispersion increases with the growing irreducible water saturation. This is result from that the pores occupied by irreducible water generate much narrower pores and more tortuous flow-paths.

In addition to the geological parameters mentioned above, the engineering parameters also have significant impact on gas mixing and CO_2-EGR performance (Dou et al. 2015; Khan et al. 2012; Khan et al. 2013). It is concluded that injecting CO_2 with a horizontal well at the lower parts, while extracting CH_4 at the upper parts of reservoirs could mitigate the breakthrough of CO_2 in the production well (Khan et al. 2012; Narinesingh and Alexander 2016). CO_2 injection during the early decline phase of natural gas production is beneficial for achieving a high CH_4 recovery, because it can ensure the displacement in supercritical phase and decrease the mixing of CO_2 and CH_4 (Khan et al. 2012). On the contrary, it may potentially lead to the trapping of CH_4 in unswept areas under the condition of high pressure. CO_2 injection in the late phase could address this issue and improve the performance of CCS, which is more attractive when CO_2-EGR as well as CO_2 storage are considered (Feather and Archer 2010; C. Khan et al. 2012). On the whole, the time of CO_2 injection corresponding the maximum incremental recovery is highly dependent on the allowable produced CO_2 concentration in the production well, which is determined by the economics of CCS projects (Regan 2010).

Whether the CO_2 is injected at the early or late stage, it is suggested to inject CO_2 at a relatively high pressure to ensure the supercritical phase in the displacement process. In this case, the distribution of CO_2 in the reservoir is dominated by gravity forces (Pooladi-Darvish et al. 2008). As the CO_2 is much denser than CH_4, CO_2 will occupy the smaller space and spread at a slower rate, which is beneficial for mitigating CO_2 breakthrough. On the contrary, if the injected CO_2 is in gas phase in the reservoir, the CO_2 will occupy a large volume and mix with the CH_4 more easily, resulting early CO_2 breakthrough (Pooladi-Darvish et al. 2008).

Regarding the injection rate, of course, high injection rate is beneficial for increasing the gas recovery (Khan et al. 2013). However, high injection rate may bring excessive gas mixing, which is harmful for natural gas production. It is suggested that the CO_2 injection rate should be lower than the CH_4 production rate to avoid the early breakthrough of CO_2 (Zangeneh et al. 2013). In Al-Hasami et al. (2005)'s study, 9% of incremental methane recovery was achieved when the CO_2 injection rate is only 13% of the production rate. Instead of injecting with constant rate, an injection scheme with a constant pressure was proposed to avoid potential risks related to high pressure (Biagi et al. 2016). The optimal injection strategy can be obtained through the optimization code based on genetic algorithm and multi-phase simulator TOUGH2 (GA-TOUGH2).

It should be pointed out that the geological parameters have a significant impact on the performance of CO_2-EGR. The parameters that affecting the viscous and gravity force, e.g., permeability, formation dip, and formation thickness, play a vital role in the stability of displacement. The fluid properties like the diffusion coefficient and water salinity are secondary parameters for affecting the CO_2 breakthrough (Regan 2010). It is worth to mention that the connate water in reservoirs has a positive impact on CO_2-EGR performance, because the dissolution of CO_2 in reservoir fluids is favorable for enhancing the storage capacity and mitigating CO_2 breakthrough in the production well (Al-Hasami et al. 2005; Patel et al. 2017; Zangeneh et al. 2013).

Several CO_2-EGR projects have been implemented around the world, including the Alberta gas field project, the K12-B field project, and the CLEAN project. The Alberta gas field project is located in

Canada. In this project, impure CO_2 with less than 2% of H_2S has been injected into the depleted Long Coulee Glauconite F gas Pool in southeastern Alberta since 2002, but the injection was terminated in 2005 result from the breakthrough of acid-gas (Pooladi-Darvish et al. 2008). The K12-B gas field with a reservoir depth of approximately 3800 m below the sea level, is located in the Dutch continental shelf in the North Sea. The reservoir fluid pressure has dropped from 40 MPa to 4 MPa with a gas recovery of 90%. The initial reservoir temperature is 128 °C (Geel et al. 2006). Over 0.1 million tons of CO that is separated from the produced gas directly at the offshore platform, has been injected into the reservoir over a period of 13 years since 2004. Monitoring data shows that the well integrity has remained stable (Vandeweijer et al. 2011). Furthermore, no major complications occurred during the lifetime of this project, demonstrating the safety of CO_2-EGR can be ensured (Vandeweijer et al. 2018). The CLEAN project was conducted between 2008 and 2011 in Germany, with CO_2 injection into the Altmark natural gas field. The risk assessment of this project has been investigated based on digital databases. The results indicated that the safety and efficiency of CO_2-EGR can be ensured. Meanwhile, the borehole integrity could be achieved without any intervention, providing a guideline on the application of CO_2-EGR (Kühn et al. 2012).

Generally, CO_2-EGR is still an immature technology. Much effort is supposed to be paid to address the problems, such as mitigating the CO_2 breakthrough and achieving favorable performance in both CH_4 production and CO_2 storage. The economic success is largely dependent on the technical maturity of CO_2-EGR and the political developments on the CO_2 emissions in the next years and decades (Kühn et al. 2012).

► **CO_2-EWR**

Similar to CO_2-EOR and CO_2-EGR, CO_2-EWR is a methodology combining CO_2 storage and saline water production (Kobos et al. 2011; Li et al. 2015; Liu et al. 2015), which is developed from the technology of CO_2 storage in saline aquifers, Fig. 1.18 shows the schematic diagram of CO_2-EWR technology. Compared with CO_2 storage in saline aquifers, there are several advantages for the operation of CO_2-EWR. Firstly, CO_2-EWR can decrease formation fluid pressure and avoid potential leakage through extracting formation water. Further, it can improve the storage capacity and achieve higher security and stability of CCS units (Liu et al. 2013). Besides, the produced saline water could be used for drinking, industrial, and agricultural utilization after desalination treatment, such as using a high efficiency reverse osmosis system (Kobos et al. 2011). Meanwhile, the deep brine resources obtained through the cascade extraction may potentially create economic profit and fill the cap of cost in the operation of CO_2 sequestration (Li et al. 2015). Kobos et al. (2011) developed a numerical simulation model to investigate the feasibility of CO_2-EWR based on a hypothetical case study related to a representative power plant and saline formation in the south-western part of the United States. In their work, the extracted saline water was treated with a high efficiency reverse osmosis system. After that, it was used as power plant cooling water. The results demonstrated that the coupled technology of CO_2 sequestration and saline water extraction and treatment is feasible for tens to hundreds of years.

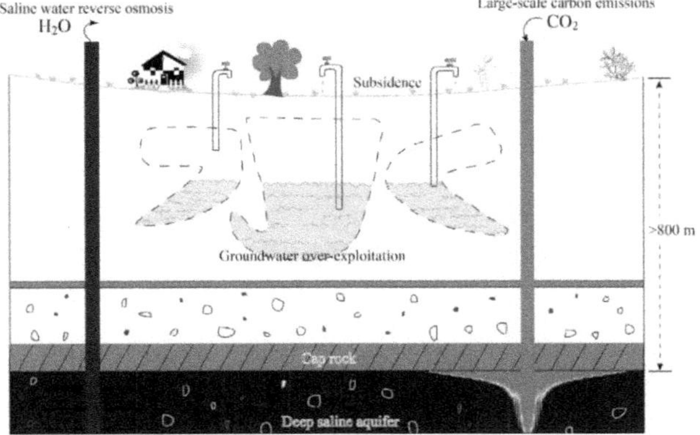

Figure 1.18 Depiction of the CO_2-EWR technology (Li et al. 2015)

Unfortunately, the added cost of extraction wells would be a shortcoming of CO_2-EWR (Kobos et al. 2011). Furthermore, the production of brine must be ceased as soon as the breakthrough of CO_2 occurs (Dewers et al. 2018). In general, the technology of CO_2-EWR may has application prospects under effective engineering design.

► **CO_2-ESGR**

In regard to the technology of CO_2-ESGR, CO_2 is injected into shale gas reservoirs to replace and displace shale gas for the purpose of enhancing the shale gas recovery, with a side benefit of CO_2 storage synchronously (Dahaghi 2010). The dominate mechanism of CO_2-ESGR is the competitive absorption of CO_2 by shale matrix (Jia et al. 2019). For example, the CO_2 sorption capacity of Muderong Shale is up to 1 mmol per gram (Busch et al. 2008). In addition, the pressure gradient displacement also plays an important role (Liu et al. 2013). Liu et al. (2013) conducted a numerical simulation to investigate the feasibility of CO_2-ESGR, and resulted that over 95% of the injected CO_2 was instantaneously adsorbed and stored in the reservoirs. However, only limited ESGR performance was detected because of the limited communication between the wells in this study. The feasibility of CO_2-ESGR on the Devonian Gas Shale Play of eastern Kentucky was analyzed by Schepers et al. (2009). They found that the Huff-and-Puff scenario was not suitable for CO_2-ESGR, while the full-field continuous CO_2 injection was a good option. A total of approximately 300 tons of CO_2 was injected within one and half months. A significantly increased gas recovery was achieved, and about half of injected CO_2 was stored. In generally, there is still a long way before the application of CO_2-ESGR, and its contribution to the mitigation of the emission of CO_2 is still limited.

▶ **CO_2-ECBR**

Regarding CO_2-ECBM, CO_2 is injected into un-mineable coal seams to displace and replace coalbed methane, simultaneously achieving CO_2 storage in the coal seams. Like the mechanism in CO_2-ESGR, CO_2 works as displace fluid and is competitive absorption by the coal seams in the process of CO_2-ECBM (Baran et al. 2014; Fang et al. 2011). The potential ECBM recovery in China is estimated to be over 3.751 Tm^3 (Yu et al. 2007), demonstrating the superiority of this technology. However, the injected CO_2 in CO_2-ECBM projects is usually less than 1 million tons per year. Further, many coal seams usually with low permeability such as those in Western Europe, which are not suitable for the application of this technology (Leung et al. 2014). Herein, the role of CO_2-ECBM on mitigating the CO_2 emissions is limited.

▶ **CO_2-EGS**

Geothermal energy is regarded as a clean, renewable, and reliable energy for its advantages of sustainability and environment friendly characteristics (Randolph and Saar 2011a). The geothermal energy is extracted through water traditionally. Brown (2000) firstly developed the concept of using supercritical CO_2 instead of water as the heat exchange fluid in EGS. It has been proven that the heat extraction efficiency of CO_2 based systems is superior to water-based systems. By using the CO_2-EGS, more regions worldwide with relatively low temperature may be used for electricity production in an economically beneficial manner (Liu et al. 2015; Randolph and Saar 2011a, 2011b). Additionally, the mobility of CO_2 is better than that of water in reservoir, which is beneficial for the production of fluids and the extraction of geothermal energy.

In recent years, the technology combining geothermal extraction and CO_2 sequestration has gained more attention (Liu et al. 2015; Randolph and Saar 2011a), because it can attain an efficient geothermal energy extraction as well as CO_2 emissions mitigation. Take the Latium Region in Central Italy for example, it is reported that the regional energy deficit could decrease by 22.1% and the CO_2 emissions could decrease by 31.3% if the CO_2-EGS was used (Procesi et al. 2013). However, CO_2-EGS is still at conceptual stage and pre-feasibility studies phase. Many efforts are supposed to be devoted towards its study before its application.

▶ **CO_2-IUL**

CO_2-IUL is a technology that leaches uranium ore out of geological formation with the help of CO_2 injection. The injected CO_2 may react with ore and minerals in the ore deposits (Wei et al. 2015). The technology of CO_2-IUL can increase the recovery of uranium and simultaneously store CO_2 in the reservoir, especially for sandstone-type uranium mining (Wei et al. 2015). However, on account of that the global annual demand of natural uranium is only around 0.1 million tons (Underhill), it would be difficult for CO_2-IUL to reduce CO_2 emission significantly because of its limits and demands.

► CH_4-CO_2 replacement from natural gas hydrates

The technology of CH_4-CO_2 replacement from natural gas hydrates (NGH) is considered as a win-win method for the exploitation of NGH and the sequestration of CO_2 in the form of hydrates formation (Chong et al. 2016; Ota et al. 2005; Yuan et al. 2012; Zhang et al. 2017). As shown in Fig. 1.19, the mechanisms of CH_4-CO_2 replacement from natural gas hydrates can be divided into four stages. Firstly, the CO_2 molecule diffuses into the surface of CH_4 hydrate and decreases the stability of CH_4 hydrate structure (Fig. 1.19a). Secondly, the CH_4 molecule escapes from the hydrate cage because of the dissociation of CH_4 hydrate (Fig. 1.19b). In the next stage, the hydrate is re-formed. As shown in Fig. 1.19c, the large cage is mainly occupied by the CO_2 molecules while the small cage is occupied by the CH_4 molecules. Finally, the CH_4 molecules diffuse from the surface of hydrate and then transform into gas. Meanwhile, the CO_2 molecules diffuse into deeper hydrate layer to continue replacing the CH_4 in hydrate (Fig. 1.19d) (Yuan et al. 2012). To improve the performance of CH_4-CO_2 replacement from natural gas hydrates, a thermal stimulation method was proposed by Zhang et al. (2017). By using this method, the CH_4 replacement exhibits an upper limit of 64.63%, and the maximum CO_2 storage efficiency can be increased up to 78.40%~96.73% (Zhang et al. 2017).

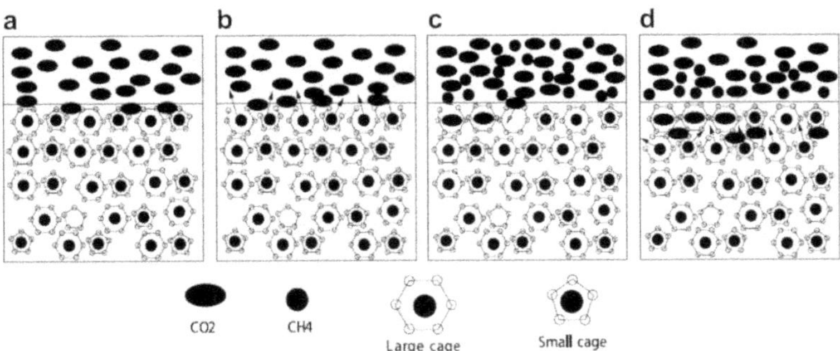

Figure 1.19 Schematic diagram for CO_2-CH_4 replacement in hydrates (Yuan et al. 2012)

Based on the concept of thermal stimulation to CO_2-CH_4 replacement, Liu et al. (2018) proposed a geothermal-assisted CO_2 replacement method (GACR). As shown in Fig. 1.20, CO_2 with ambient temperature was injected into geothermal reservoir for heating. Subsequently, the heated CO_2 flows upward into the hydrate bearing layer (HBL) to enhance the NGH dissociation. Numerical simulation results indicated that the GACR method can significantly accelerate the dissociation of NGH and increase the recovery of CH_4. However, it should be mentioned that the application of this method is limited by the strict condition required, i.e., a thermal reservoir located below the methane hydrate reservoir.

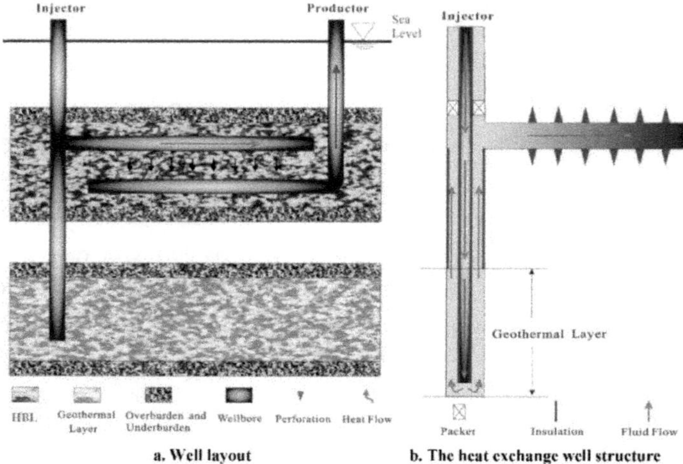

Figure 1.20 Schematic well group configuration diagram of GACR (Liu et al. 2018)

The technology of CH_4-CO_2 replacement from natural gas hydrates is currently in the preliminary experimental and numerical study phase (Lee et al. 2017; Xu et al. 2018; Zhang et al. 2017). However, it is expected that great progress will be made in the near decades under the stimulation of methane hydrate production and CO_2 storage.

In a summary of this part, the utilization of CO_2 for resources production and CO_2 sequestration are suggested to be designed for the whole process of engineering operation to co-optimize their performance. It is worth to mention that the technology of artificial intelligence may would be helpful for the optimization. However, despite the technology of CO_2-EOR has been used commercially, the other technologies are still in the pilot plant phase in terms of technology readiness level (Bui et al. 2018). In the next few years, CO_2-EOR would be the most promising technology that combining the utilization and storage of CO_2. It should be mentioned that CO_2-EGR is another promising technology, whose application is limited by the breakthrough of CO_2 and the mixing of CO_2 with CH_4. Considering that the mixing behavior is significantly affected by the geological parameters, i.e., permeability, porosity, residual water saturation, and engineering parameters, i.e., injection pressure, injection rate, production rate, a site selection system for CO_2-EGR project is encouraged to be developed.

1.5.2 Co-injection of CO_2 with impurities

The biggest obstacle for the application of large-scale CO_2 storage is the lack of financial incentives (Gislason and Oelkers 2014). Most strategies of CO_2 storage cannot generate economical profit, thus the measures to cut down the cost of CO_2 storage is very important and beneficial for the application of this technology in large-scale. It is reported that the overall cost of carbon capture and storage is dominated by the process of capture and gas separation, which costs \$55 to \$112 per ton of CO_2 (Gislason and

Oelkers 2014). Therefore, co-injection of CO_2 with impurities can be a cost-effective option for CO_2 storage because it can significantly reduce the cost on gas separation (Jafari Raad and Hassanzadeh 2017).

There are several impurity gases may be co-injected with CO_2 into underground such as CH_4, H_2S, SO_2, N_2, and O_2. Among them, H_2S and CH_4 are usually mixed with CO_2 in produced acid gas, which can be used in CO_2 sequestration. Other gases are the main components of flue gas which is captured from the major CO_2 emission sources such as power plants (Wang et al. 2012).

It should be mentioned that the impure CO_2 can also be utilized especially for CO_2-EOR with multicontact miscible CO_2 flooding (Metcalfe 1982). The MMP is a key control variable in the multicontact miscible displacement procedure because it has notable impact on the design and development of assets. In addition, the MMP is closely related to the economically feasibility of the CO_2-EOR operation. A low MMP is supposed to be favorable for CO_2-EOR, while a high MMP is disadvantageous for CO_2-EOR operation and may increase the risk for the fracturing of formation because of the higher injection pressure required. Generally, the impurities in CO_2 would affect the MMP a lot. Specifically, the presence of H_2S and SO_2 in CO_2 would reduce the MMP (Metcalfe 1982; Sayegh et al. 1987), while the presence of CH_4, N_2, and O_2 would increase the MMP of CO_2 (Yang et al. 2007; Zhang et al. 2004).

The presence of impurities may also change the thermophysical properties and phase behavior of impurity-laden CO_2 (Martha Hajiw et al. 2018), and further affect the performance of CCS. For example, N_2 would lead to a delay of CO_2 breakthrough when it is co-injected, this is result from that the solubility of N_2 in irreducible water is much lower than that of CO_2 (Turta et al. 2007). However, the N_2 would decrease the density of the dissolved phase and increase the risk of the CCS unit in the long term (Li et al. 2015). Generally, the storage capacity of reservoirs decreases proportionally to the concentration and the compressibility factor of impurities (Barrufet et al. 2010). It should be mentioned that the reduced storage capacity may be even higher than the volume fraction of impurities when O_2 is co-injected. But, the negative impact of impurities on the storage capacity can be alleviated by storing the impure CO_2 in a reservoir with high temperature (Wang et al. 2012). This is result from that the impact of impurity gas on the density of impurity-laden CO_2 is minor in the condition of high temperature.

The impact of N_2 on the performance of CO_2 sequestration has been investigated extensively by researchers. Especially, the impact on the storage capacity, solubility trapping, and the CO_2 plume migration has been paid for attention. Barrufet et al. (2010) analyzed the effect of N_2 as impurities in the CO_2 stream on the storage capacity. Their results indicated that the CO_2 storage capacity decreases proportionally to the concentration of N_2 in injected gases. Li and Jiang (2014, 2020) and Li et al. (2015) conducted numerical simulations to analyze the effect of N_2 on the dissolution trapping mechanism and the onset of convection in a water saturated porous media. Their results showed that a density reduction of the aqueous phase was found when the N_2 was co-injected with CO_2 into the formation water, resulting the delay of the onset of convection. It should be mentioned that the delay of the convection onset is not favored for accelerating the dissolution rate of CO_2. Regarding this issue, Mahmoodpour et al. (2018) conducted theoretical and experimental studies based on a high pressure Hele-Shaw cell and resulted that the onset of convection is dependent on the mole fraction of N_2. Especially, the onset of convection for the CO_2 containing 20% N_2 was close to that of pure CO_2. Wei et al. (2013) performed a numerical study on the co-injection of N_2 and CO_2 into an aquifer by using COMSOL-Multiphysics software. They resulted that the occurrence of N_2 would decreases the storage capacity of CO_2. In addition, the chromatographic partitioning phenomenon was observed, indicating the potential application of N_2 as a non-toxicity monitoring gas for the leakage of CO_2 in CCS units. The chromatographic partitioning phenomenon was also observed by Li and Jiang (2017) and Park et al. (2017) in the simulation of co-injection of CO_2 with N_2 into saline aquifers. Wu et al. (2017) conducted experiments to investigate the impact of N_2 on the saturation of gas mixture by using X-ray CT scanning. Their results demonstrated that the saturation increases with flow rate for the scenario with 80% CO_2, while the trend is opposite for the scenario with 95% CO_2, showing the complicated mechanisms of impure CO_2 saturation affected by the flow rate and gas proportion.

The impact of O_2 on the performance of CO_2 storage was studied by Nicot et al. (2013). Their results resulted that the CO_2 storage capacity and injectivity, which is characterized by the mixture density and the ratio of density and viscosity, decrease at shallow formation while recover with increasing depth. In addition, the CO_2 plume extent is greater than that of pure CO_2 when O_2 is co-injected, which is favorable for solubility trapping. The impact of O_2 on the convective mixing of CO_2 storage in saline aquifers was investigated by Li and Jiang (2020). Their results showed that O_2 would lead to delay onset and weakened convection, which is disadvantageous for accelerating the solubility trapping of CO_2. Due to the chemical reactivity of O_2 in the reservoir conditions, some batch experiments have been performed to investigate the impact of O_2 on the reaction of rock minerals (Jung et al. 2013; Nicot et al. 2013). The general finding is the production of iron oxides by pyrite oxidation caused by the presence of O_2, which was observed in the experimental studies associated with both sandstone and shale caprock (Jung et al. 2013; Nicot et al. 2013).

The effect of H_2S with a fraction of less than 30% as impurity gas on the dissolution of CO_2 is not significant (Jafari Raad and Hassanzadeh 2017). However, when it was co-injected with CO_2 under the condition of 20 MPa and 45 °C, the H_2S with a concentration over 20% has a potential to decrease the interfacial tension and increase the contact angle, resulting a low capillary force (Chen et al. 2017). This means that H_2S may increase the risks associated with gas leakage, which should be paid attention to. It is worth to mention that the impure CO_2 with H_2S can be trapped by hematite even in a dry system driven by the reduction of ferric iron in hematite by sulfide species, which verifies the feasibility of co-injection of CO_2 with H_2S (Alpermann et al. 2016). Zhang et al. (2020) conducted a numerical study by using the TMVR_EOSG module based on the TOUGHREACT code to investigate the feasibility of co-injection of H_2S with CO_2 into carbonate reservoirs in the Tarim Basin. Their results showed that the migration of the H_2S is slower in both gas and aqueous phases compared with CO_2 due to the difference of solubility, which provide an indicator for the potential breakthrough of the injected acid gas. They also concluded that the pressure buildup is below the allowable pressure increment, and the porosity change is negligible, demonstrate the feasibility of co-injecting H_2S with CO_2 in the carbonate formation.

The CH_4 produced from acid gas reservoir may also serve as impurity gas and be injected into underground in CCS projects. Though the concentration of CH_4 in the injected gases up to 20%, there is no significant negative impact on the interfacial tension and wettability (Chen et al. 2017). However, the storage capacity of reservoirs decreases proportionally to the concentration and compressibility factor of CH_4 (Barrufet et al. 2010). Due to the concentration of CH_4 in injected CO_2 is very low, its effect on the CCS is minor and neglectable.

The presence of SO_2 controls the acid-induced reactions with calcium-rich minerals when it is co-injected into reservoirs as an impurity gas. However, the quantitative effect is usually very minor and can be neglected for sandstone such as the German Bunter Sandstone (Fischer et al. 2018). Generally, the porosity in sandstone would increase under the impact of SO_2, while the porosity of shale layer would decrease due to the conversion of dominant calcite to anhydrite (Wolf et al. 2016). For instance, the conversion of Ca^{2+} bearing carbonate to anhydrite is observed when SO_2 was co-injected with CO_2 into the German Bunter Sandstone (Wolf et al. 2017). A field experiment was conducted by Vu et al. (2018) to study the geochemical impacts of SO_2 and O_2 as impurities on the reactions of minerals and fluids in a siliciclastic reservoir. During this experiment, the CO_2-saturated water with impurities was injected into reservoirs and allowed to interact with minerals for three weeks. Their results indicated that the pyrite dissolved because the O_2 acting as an oxidizing agent. It should be mentioned that the concentration of SO_2 and O_2 are 67 ppm and 6150 ppm respectively, which is too low to cause a significant effect on the fluid-rock interaction. It can be inferred that the effect of impurities on the interaction with formation rock is highly dependent on the composition of minerals, which should be analyzed site specifically. It is worth to mention that co-injection of SO_2 and CO_2 could suppress the Joule-Thomson cooling effect, which is a beneficial thermal consequence for CCS (Ziabakhsh-Ganji and Kooi 2014a).

There are some researches focus on the influence of the mixture of N_2 and O_2 on the performance of CCS. Wang et al. (2012) investigated the impact of N_2 and O_2 on the storage capacity of CO_2. Their results indicated that the reduction of structural trapping capacity of CO_2 caused by the impurities is achieved by reducing the density of CO_2. Moreover, the reduction of the trapping capacity is greater than the fraction of the impurity gas. Lei et al. (2016) conducted a numerical simulation for co-injection of N_2 and O_2 with CO_2 into the aquifers located at the Tongliao CCS site. They resulted that the impact of the ratio of N_2 and O_2 on the distribution pattern of gas in the reservoir is not significant, while it affects the mole fraction and the transport distance of the gas mixture. Additionally, the chromatographic partitioning phenomenon was observed in both gas and aqueous phases in their work. This phenomenon was verified by the pilot field experiment conducted by Wei et al. (2015). During the operation of this field experiment, a total of 200 tonnes of CO_2 and 30 tonnes of air (N_2 and O_2) were injected into the aquifer. However, the air may not be representative for the flue gas that is usually co-injected with CO_2 for CCS because of the difference of gas mole fraction. In addition, the temperature and the initial fluid pressure in the aquifer is 15 °C and 2.1 MPa respectively, in which condition the injected CO_2 is in gaseous state instead of supercritical state. Therefore, it cannot characterize the main state of CO_2 in the scenarios of CCS. Generally, the forementioned research work laid a foundation for the co-injection of N_2 and O_2 with CO_2 for CCS. However, to the best of our knowledge, almost all of the storage sites considered for injecting impurities with CO_2 are deep saline aquifers, whereas the depleted gas reservoirs have not been included.

In short, co-injection of CO_2 with impurity gas is an effective strategy to reduce the overall cost of CCS projects. It should be mentioned that the impact of impurities on the thermophysical properties of reservoir fluids, and the chemical interaction between different impurities and formation rocks need to be further studied, to reduce the uncertainties in the design and operation of CCS projects.

1.5.3 CO_2 as cushion gas for underground natural gas storage in depleted reservoirs

Natural gas storage is principally used for meeting the widely fluctuating demand of natural gas, especially the high peak demands in winter (Katz and Tek 1981; Demirel et al. 2017; Tek 1989). In addition, the development of natural gas storage has been further promoted by a profitable business model, i.e., storing the gas at a lower price and selling it at a higher price based on demand loads (Speight 2007). There are several important strategies for natural gas storage such as gas tanks, salt caverns, and gas pipeline (Wang and Economides 2009). Compared with the forementioned gas storage strategies, underground gas storage reservoirs (UGSR) have the advantages of large storage capacity and superior economy. Therefore, UGSR has been widely used in North America and Europe, and has also gained increasing attention in China (Evans 2009; Juez-Larré et al. 2016; Laier 2012; Teatini et al. 2011; Wang and Economides 2012; Zhang et al. 2017). It should be mentioned that the underground gas storage is mostly a seasonal operation that usually been charged in summer and discharged in winter. During the operation process, cushion gas is vital to maintain suitable reservoir pressure as well as keep a stable production operation.

In the case of UGSR based on depleted gas reservoirs, the native natural gas is commonly applied as the cushioning gas (Oldenburg 2003). There are some shortcomings regarding this gas storage method. Firstly, it may lead to a great deal of waste because approximately 40%-70% of stored gas works as cushion gas to provide pressure support and cannot be extracted (Misra et al. 1988). Secondly, the security of the UGSR during CH_4 injection may be threatened dramatically by the increased reservoir pressure because of the relatively low compressibility of the cushion gas. To address these issues, utilizing supercritical fluids such as supercritical CO_2 as cushion gas is a potential strategy. The physicochemical properties such as density and compressibility exhibit anomalous behavior (Imre et al. 2014; Raju et al. 2017; Velmovszki et al. 2019) when the CO_2 is in the under the critical conditions. It should be mentioned that the critical condition exactly can be achieved in the storage formations, demonstrating the potential prospects for the utilization of CO_2 as cushion gas in the natural gas storage reservoir. In comparison with the native gas cushion, an incremental of more than 30% CH_4 can be stored with CO_2 gas cushion (Oldenburg 2003). Further, it may bring an additional benefit result from enhancing gas recovery in the post-CO_2 storage and enhanced gas recovery (CSEGR) process for a depleted gas reservoir. However, as shown in Fig. 1.21, the mixing of CO_2 and CH_4 with the injection and production of working gas could be an obstacle for the utilization of CO_2 as cushion gas (Oldenburg 2003), which should be paid for serious attention.

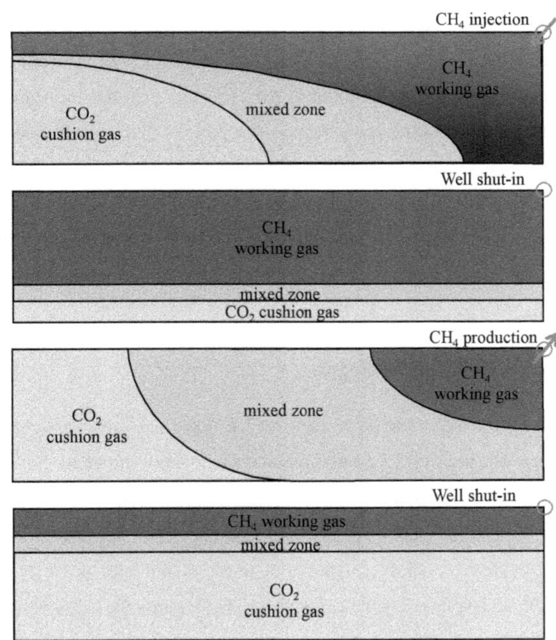

Figure 1.21 The schematic diagram of natural gas storage reservoir with CO_2 cushion gas (Cao et al. 2020)

Some research efforts have been paid to investigate the mixing behavior of CO_2 and CH_4 in UGSR. Oldenburg (2003) studied a numerical simulation by using a two-dimensional reservoir model, to demonstrate the suitability of utilizing CO_2 as cushion gas in natural gas storage. Based on this reservoir model, Ma et al. (2019) analyzed the impact of geological parameters on the gas mixing behavior in UGSR with CO_2-based cushion gas by using a hydromechanically coupled model. Their results indicated that the mixing region decreases with increasing reservoir thickness and dip angle. However, it cannot provide the guidelines for utilizing CO_2 as cushion gas for UGSR in depleted gas reservoirs because of the following limits in the studies. Firstly, the reservoir model was initially saturated with CO_2 which neglects the replacing process of the native CH_4 by the injected CO_2 during its injection into the depleted gas reservoir. In addition, it did not characterize the seasonal injection and production process, as the mixing behavior has been studied only within 180 days. Moreover, the presence of residual water in reservoir that could affect the mixing behavior, has not been considered in their work. Niu and Tan (2014) analyzed the effects of reservoir porosity and initial operation pressure on the mixing behavior of the cushion and working gases based on the three-dimensional gas-water two phase theory. Their results indicated that the reservoir with high porosity and large initial pressure is beneficial for limiting the formation and migration of the mixed zone. Oldenburg and Pan (2013) conducted a one-dimensional radial simulation to analyze the pressurization and gas-gas mixing behavior. It was found that the pressure increment during working gas injection reduces if CO_2 is applied as cushion gas. They also discovered that the impact of the CO_2 cushion on the reservoir pressure during the production period is much more significant than the one during the injection period. This is result from that the production rate is higher than the injection rate in engineering operation. The behavior of CO_2 and N_2 as cushion gases were compared by Kim et al. (2015), the results indicated that CO_2 is more suitable as cushion gas in depleted gas reservoirs in terms of the productivity index. However, in these works, a single well for both cushion gas injection and working gas production is very likely to accelerate the mixing of gases in UGSR, which is harmful in the engineering field.

1.6 Prospects of CCS/CCUS technologies

The economic factor for the CCS projects is considered as one of the most important incentives for the application in industry. The price for CO_2 emissions at the first major carbon market and also the biggest one, i.e., the European Union Emission Trading System, is approximately $7 per ton of CO_2, which is much cheaper than the cost of CCS (Gislason and Oelkers 2014). Therefore, there is no financial incentive for the CCS industries unless a higher price of carbon emission is set. This demonstrates the important role that should be played by the government in mitigating the CO_2 emissions.

Tab. 1.6 shows the large-scale CCS projects (more than 0.4 Mtpa) throughout the world until the 2020's. It can be seen that the CCUS for EOR and CCS in saline formations make major contributions towards CO_2 storage, which is in accordance with the prediction results obtained by Mac Dowell et al. (2017). Nearly half of 51 large scale CCS projects scheduled are designed for EOR, demonstrating the economic viability in EOR operations. In addition, a total of 21 projects related to CCS in saline formations are also planned because its CO_2 storage capacity may up to 4 Mtpa. It should be mentioned that the average CO_2 storage capacity of CCS in depleted gas fields is much greater than that of EOR and may reach to

2.8 Mtpa, indicating the prospects on the mitigation of CO_2 emissions. However, due to the great extent of the mixing between CO_2 and CH_4 which is an obstacle for enhancing additional recovery of CH_4, there are only 3 large-scale CCS projects in depleted gas fields scheduled in near future. With the development of technology to address this issue, it can be inferred that CCS in depleted gas fields can play a more important role in mitigating the CO_2 emissions.

Table 1.6 Large-scale CCS projects (more than 0.4 Mtpa) throughout the world until the 2020's (Cao et al. 2020)

Strategy	EOR	CCS with potential EOR	CCS		Total
			Saline formation	Depleted gas fields	
Quality of project	24	3	21	3	51
CO_2 capture capacity (Mtpa)	42.11~43.41	8.1~8.6	40.35~85.1	7.5~8.5	98.06~145.61
Average CO_2 capture capacity (Mtpa)	1.75~1.81	2.7~2.87	1.92~4.05	2.5~2.83	1.92~2.86

In summary, the status of the strategies for CO_2 storage has been discussed in view of assessing the risks as well as improving the cost-effectiveness. In addition, the role of CCS technologies and their potential contribution on mitigating the CO_2 emissions in future were summarized. Based on the studies carried out in aforementioned literature review, the following results have been obtained.

Firstly, sequestration of CO_2 in depleted oil and gas reservoirs could play an important role in mitigating CO_2 emissions in near future, because the existing installed equipment and comprehensively characterized reservoir integrity will significantly reduce the cost of CCS. The leakage of CO_2 through abandoned wells may be an obstacle for the application of this technology. To address this issue, the long-term experiments and molecular dynamic simulations are supposed to be conducted to figure out the kinetics between CO_2 with the well string, cement, as well as formation rock minerals under the relevant conditions. Secondly, if implemented on a large scale, CO_2 storage in saline aquifers may make the biggest contribution in reducing CO_2 emissions because of its huge storage capacity. Moreover, the scientifically proven technologies such as CO_2 storage in coal beds, deep ocean, and deep-sea sediments are still immature technologies and do not appear to be capable of making a great contribution to the mitigation of CO_2 emissions in the foreseeable future.

Another point is the requirement to investigate accurate risk assessment associated with CO_2 storage and provide a guideline for the design and construction of CCS projects. Attempting to make the CCS assessment more intelligential, the machine learning technology is suggested to be used.

It has also been demonstrated that the direct mineralization of flue gas by coal fly ash would be a promising technology. This is result from that the flue gas could be mineralized directly without separation. Also, the pH swing mineralization may be another promising technology for the CO_2 storage because of its sustainability. The recyclable and cheap chemical reagents are encouraged to be developed and introduced into the mineralization process to make this technology more cost-effectiveness.

Among the variety of CCUS technologies, CO_2-EOR followed by CO_2-EGR is supposed to play the most important role in mitigating CO_2 emissions in the next few years. The utilization of other technologies seems to be negligible in near future. Co-injection of impurities with CO_2 is an effective methodology to reduce the overall cost of CO_2 storage. It should be pointed out that the physical and chemical effects of the impurities on reservoir fluids and formation rock should be studied site specific, to reduce the uncertainties in CO_2 storage.

The government is supposed to play an important role in mitigating CO_2 emission. A higher tax on CO_2 emissions and financial subsidy on CO_2 storage is encouraged to accelerate the deployment of CCS projects on large-scale.

1.7 Research objectives

According to the aforementioned discussion, there are two key factors limit the application of CCS in large-scale. The first one is the risks associated with CCS. The second one is the lack of financial incentives of CCS operation. This thesis attempts to focus on these two problems and provide some strategies to address them.

Regarding the risks associated with CCS, this thesis conducted a parametric uncertainty analysis for CO_2 storage and determined the general role of different geomechanical and hydrogeological parameters in response to CO_2 injection. It can be useful to guide time and effort spent in mitigating the uncertainties in these parameters to acquire trustworthy model forecasts and risk assessments in CCS projects.

Regarding the financial incentives of CCS operation, this thesis attempts to increase the cost-effectiveness of CCS through the following two strategies. First, considering that the processes of carbon capture and gas separation dominate the overall cost of CCS, this thesis investigated the suitability of co-injecting CO_2 with N_2 and O_2 that are the main impurities from flue gas into depleted gas reservoirs. It should be mentioned that the options of dedicated for CO_2 storage and CSEGR with the impure CO_2 were considered respectively, which may potentially decrease the overall cost and accelerate the application of CCS in large-scale. Second, the suitability of utilizing CO_2 as cushion gas for UGSR in depleted gas reservoir was studied, which is also beneficial for improving the cost-effectiveness of CO_2 storage.

1.8 Thesis outline

This thesis focuses on the related risks and cost-effectiveness of CCS, the numerical study of underground CO_2 storage and the utilization in depleted gas reservoirs were conducted. The research contents and flowchart of this thesis are shown in Fig. 1.22.

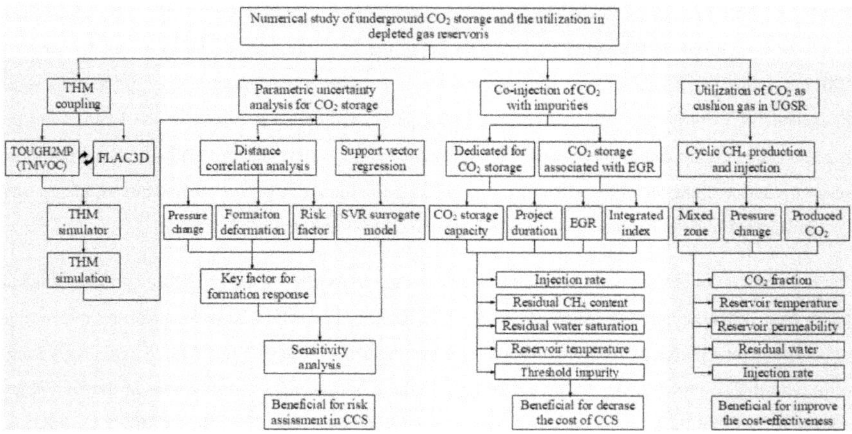

Figure 1.22 Research content and flowchart

The contents of this thesis are divided into five chapters. In chapter 2, the theoretical background of the coupled THM processes of CO_2 storage is introduced. Firstly, the mass and energy transport during the multiphase and multicomponent flow are presented. Secondly, the governing equations of the mechanical behavior for formation rocks such as the constitutive equations are introduced. Thirdly, the mechanisms of TH, TM, and HM coupling are presented respectively. Thereafter, the software of TOUGH2MP and FLAC3D, as well as the coupling of TOUGH2MP (TMVOC)-FLAC3D are introduced.

Based on the developed THM simulator TOUGH2MP (TMVOC)-FLAC3D, the parametric uncertainty analysis for CO_2 storage is conducted in chapter 3. In this chapter, a 2D model with geometric parameters based on the worldwide large-scale CCS projects is developed firstly. Then the geomechanical and hydrogeological parameters, including Young's modulus, Poisson's ratio, Biot's coefficient, permeability, permeability anisotropy ratio, and the injection rate are sampled randomly based on the Quasi-Monte Carlo method. Further, numerical simulation is performed on every data point by using the developed TOUGH2MP(TMVOC)-FLAC3D simulator. Based on the simulation results, the correlation between different parameters and the pressure change, formation deformation, and the risk factor are analyzed respectively by using the distance correlation (Székely, Rizzo, & Bakirov, 2007), which is beneficial for determining the key parameters for different formation responses. Moreover, the support vector regression (SVR) surrogate models are developed based on the machine learning approach in SVR (Chang, 2011; Drucker, 1997; Vapnik, 2013). The SVR surrogate model are verified by the numerical simulation results. Lastly, the sensitivity analysis of the key parameters on the formation response, including pressure change and formation deformation is conducted by using the trained SVR surrogate models.

For the purpose of decrease the cost of CCS, co-injection of CO_2 with impurity gas into depleted gas reservoir is investigated in Chapter 4. The main impurities from the flue gas that is the main source of CO_2 emissions, i.e. N_2 and O_2, are considered with three different concentrations. Both the purpose of dedicated for CO_2 storage and CSEGR are investigated. Further, the impact of the key geological parameters of the depleted gas reservoir including the residual CH_4 content, residual water saturation, reservoir temperature, and the engineering parameters including the injection rate and threshold impurity concentration on the performance of CO_2 storage and CSEGR are analyzed detailly.

In chapter 5, the suitability of utilizing CO_2 as cushion gas in UGSR is analyzed based on the geological parameters of the Donghae depleted gas reservoir in Korea. The cyclic CH_4 production and injection is conducted over a period of 15 years to acquire the mixing behavior of CO_2 and CH_4 in a relatively long-term period. The mixed zone, pressure change, and the produced CO_2 concentration in the UGSR are considered as the indicator of the performance of CO_2 cushion gas. The effect of the CO_2 fraction, reservoir temperature, reservoir permeability, residual water, and production rate on the mixing behavior and gas production in the UGSR are discussed in detail.

2 THM coupled geo-processes of underground CO_2 storage

2.1 Multiphase and multicomponent flow

2.1.1 Mass transport

► **Basic equation**

For the multiphase and multicomponent flow in a system, the mass change in the pore space is equal to the sum of the injection source and the flux from the boundaries, which can be expressed as (Pruess et al. 1999):

$$\frac{d}{dt}\int_{V_n} M^k dV_n = \int_{\Gamma_n} \mathbf{F}^k \cdot \mathbf{n} d\mathbf{\Gamma}_n + \int_{V_n} q^k dV_n \tag{2.1}$$

where V_n denotes the control volume for an arbitrary subdomain in the flow system [m³], $\mathbf{\Gamma}_n$ represents the closed surface of V_n [m²], M^k denotes the mass per volume [mol/m³], k is the mass components, $\mathbf{F}^k$ denotes the mass flux [mol/(m²·s)], and q^k is the sinks and sources [mol/(m³·s)].

It is assumed that all components and phases distribute uniformly in the element in porous media. The mass accumulation term M^k of component k for mass transport can be obtained as its sum in each phase β:

$$M^k = \phi \sum_{\beta} S_\beta \rho_\beta X_\beta^k \tag{2.2}$$

where ϕ is the porosity [-], S_β is the saturation of phase β [-], ρ_β is the density of phase β [kg/m³], and X_β^k is the mass fraction of component k in phase β [-].

There are three mechanisms dominate the mass transport in the process of multiphase and multicomponent flow, including the advection, diffusion, and dispersion.

► **Advection**

The total advective mass flux $\mathbf{F}^k$ of component k can be calculated as the sum of the individual flux in every phase according to its phase fraction X_β^k, as follows:

$$\mathbf{F}^k = \sum_{\beta} \mathbf{X}_\beta^k \mathbf{F}_\beta \tag{2.3}$$

Then the mass flux $\mathbf{F}_\beta$ in phase β can be obtained by using the Darcy's law:

$$\mathbf{F}_\beta = \rho_\beta \mathbf{u}_\beta = -k \frac{\mathrm{k}_{r\beta} \rho_\beta}{\mu_\beta} \left(\nabla P_\beta - \rho_\beta \mathbf{g} \right) \tag{2.4}$$

where $\mathbf{u}_\beta$ is the Darcy velocity for phase β [m/s], k and $k_{r\beta}$ are the absolute permeability and relative permeability of phase β, respectively [m²], μ_β is the viscosity of phase β [Pa·s], P_β is the fluid pressure of phase β [Pa], and ρ_β is the mass density of phase β [kg/m³].

The absolute permeability of gas at low pressure can be calculated according to the work of Klinkenberg (1941) as expressed in Eq. 2.5:

$$k = k_\infty \left(1 + \frac{b}{P}\right) \tag{2.5}$$

where k_∞ denotes the permeability at "infinite" pressure, b presents the Klinkenberg parameter.

The relative permeability is a function of the phase saturation. There are several empirical models can be used to calculate the relative permeability. Regarding two phase (gas and liquid) flow, the Corey (1954)'s function and Van Genuchten (1980)'s function has been wildly used to characterize the flowing behavior. The Corey (1954)'s function is expressed as follows:

$$k_{rl} = \hat{S}^4 \tag{2.6}$$

$$k_{rg} = (1-\hat{S})^2\left(1-\hat{S}^2\right) \tag{2.7}$$

where $\hat{S} = (S_l - S_{lr})/(1 - S_{lr} - S_{gr})$, S_{lr} represents the residual liquid saturation [-], S_{gr} is the residual gas saturation [-].

The Van Genuchten (1980)'s function can be written as:

$$k_{rl} = \sqrt{S^*}\left\{1-\left(1-\left[S^*\right]^{1/\lambda}\right)^{\lambda}\right\}^2 \tag{2.8}$$

where $S^* = (S_l - S_{lr})/(1 - S_{lr})$, λ is the exponent [-]. The gas relative permeability can be calculated according to the Corey (1954)'s function.

The P_β in Eq. 2.4 can be obtained by:

$$P_\beta = P + P_{c\beta} \tag{2.9}$$

where P is the pressure of reference phase [Pa], $P_{c\beta}$ is the capillary pressure [Pa].

The capillary pressure can be calculated by the empirical functions such as Van Genuchten (1980)'s function:

$$P_{cap} = -P_0\left[\left(S^*\right)^{-1/m} - 1\right]^{1-m} \tag{2.10}$$

where P_0 is the strength coefficient [Pa] and m is the van Genuchten's shape parameter [-].

► **Diffusion**

Molecular diffusion may play an important role in the process of mass transport, especially when the advective velocity is small. Diffusion flux is proportional to the gradient of concentration and can be expressed according to the Fick's low as follows:

$$\mathbf{f} = -\mathrm{d}\nabla \mathrm{C} \tag{2.11}$$

where d denotes the effective diffusivity, which is dependent on the porous media, the pore fluid, and the diffusing component.

The concentration can be expressed by the various models such as De Marsily (1986)'s function. Regarding the multiple components diffuse in a multiphase flow system, the complex non-linear behavior occurs result from that all the concentration variables may affect the effective diffusivities. In addition, the phase saturation dependent tortuosity from the coupling of advective and diffusive transport is another non-linear effect. Therefore, it is not possible to present a model to accurately characterize the multiphase diffusion under all circumstances. Pruess et al. (1999) developed a pragmatic model to characterize the diffusive flux of component k in phase β (gas and liquid), which can be expressed as:

$$\mathbf{f}_\beta^k = -\phi\tau_0\tau_\beta\rho_\beta \mathrm{d}_\beta^k \nabla \mathrm{X}_\beta^k \tag{2.12}$$

where ϕ denotes the porosity [-], τ_0 is a factor dependent on the porous media [-], τ_β represents the coefficient that depends on the saturation of phase β [-], thus $\tau_0\tau_\beta$ denotes the tortuosity [-]. ρ_β is the density [kg/m^3], d_β^k represents the molecular diffusion coefficient for component k in phase β [m^2/s].

Therefore, a single effective multiphase diffusion coefficient which combines all material constants and tortuosity factors can be defined as:

$$\Sigma_\beta^k = \phi\tau_0\tau_\beta\rho_\beta \mathrm{d}_\beta^k \tag{2.13}$$

For the two-phase conditions, the diffusive flux can be expressed as:

$$\mathbf{f}^k = -\Sigma_1^k \nabla X_1^k - \Sigma_g^k \nabla X_g^k \tag{2.14}$$

► **Dispersion**

The mass transport by hydrodynamic dispersion can be expressed as:

$$\overline{\mathbf{D}}_{\beta}^{k} = D_{\beta,T}^{k}\bar{\mathbf{I}} + \frac{\left(D_{\beta,L}^{k} - D_{\beta,T}^{k}\right)}{u_{\beta}^{2}}\mathbf{u}_{\beta}\mathbf{u}_{\beta} \tag{2.15}$$

where $D_{\beta,\mathrm{L}}^{k}$ is the longitudinal dispersion coefficient [m^2/s], $D_{\beta,T}^{k}$ is the transverse dispersion coefficient [m^2/s], which can be expressed as:

$$D_{\beta,\mathrm{L}}^{k} = \phi\tau_0\tau_\beta \mathrm{d}_{\beta}^{k} + \alpha_{\beta,L}\mathrm{u}_{\beta} \tag{2.16}$$

$$D_{\beta,T}^{k} = \phi\tau_0\tau_\beta d_{\beta}^{k} + \alpha_{\beta,T}\mathrm{u}_{\beta} \tag{2.17}$$

where τ_0 is a factor dependent on the porous media [-], τ_β represents the coefficient that depends on the saturation of phase β [-], thus $\tau_0\tau_\beta$ denotes the tortuosity [-]. d_{β}^{k} is the molecular diffusion coefficient for component k in phase β [m^2/s], α_L is the longitudinal dispersivity [m], α_T is the transverse dispersivity [m].

2.1.2 Energy transport

The energy flux consists conductive and convective components, which can be expressed as:

$$\mathbf{F}^{\mathrm{NK}+1} = -\lambda\nabla\mathrm{T} + \sum_{\beta} h_\beta \mathbf{F}_\beta \tag{2.18}$$

where $\nabla\mathrm{T}$ is the temperature gradient K/m, λ denotes the thermal conductivity W/(m·K), h_β is the specific enthalpy in phase β [J/kg].

2.2 Mechanical behavior of formation rocks

2.2.1 Stress and strain

The stress and strain are second-order tensors and can be expressed as a 3 × 3 matrix. There are three principal directions that are fixed, i.e., it will not change with the varies of axis orientations. There are three principal values in the stress and strain tensors (Fig. 2.1), which can be obtained as follows:

$$\left|\sigma_{ij} - \lambda\delta_{ij}\right| = -\lambda^3 + I_1\lambda^2 - I_2\lambda + I_3 = 0 \tag{2.19}$$

where I_1, I_2, and I_3 are three invariants, which are independent on the coordinates and can be given as follows:

$$I_1 = \sigma_{ii} = \sigma_{xx} + \sigma_{yy} + \sigma_{zz} \tag{2.20}$$

$$I_2 = \frac{1}{2}\left(\sigma_{ii}\sigma_{jj} - \sigma_{ij}\sigma_{ji}\right) = \sigma_{xx}\sigma_{yy} + \sigma_{yy}\sigma_{zz} + \sigma_{xx}\sigma_{zz} - \tau_{xy}^2 - \tau_{yz}^2 - \tau_{xz}^2 \tag{2.21}$$

$$I_3 = \det(\sigma_{ij}) = \sigma_{xx}\sigma_{yy}\sigma_{zz} + 2\tau_{xy}\tau_{yz}\tau_{xz} - \tau_{xy}^2\sigma_{zz} - \tau_{yz}^2\sigma_{xx} - \tau_{xz}^2\sigma_{yy} \tag{2.22}$$

Based on the first invariant, the mean stress and volumetric strain are defined as follows:

$$\sigma_m = \frac{\sigma_{ji}}{3} \tag{2.23}$$

$$\varepsilon_{x_n} = \varepsilon_{ij} \tag{2.24}$$

By using the mean stress and volumetric strain, the stress and strain can be written as the sum of isotropic part and deviatoric part as follows:

$$\sigma_{ij} = \sigma_m \delta_{ij} + s_{ij} \tag{2.25}$$

$$\varepsilon_{ij} = \varepsilon_{vol}\delta_{ij} + e_{ij} \tag{2.26}$$

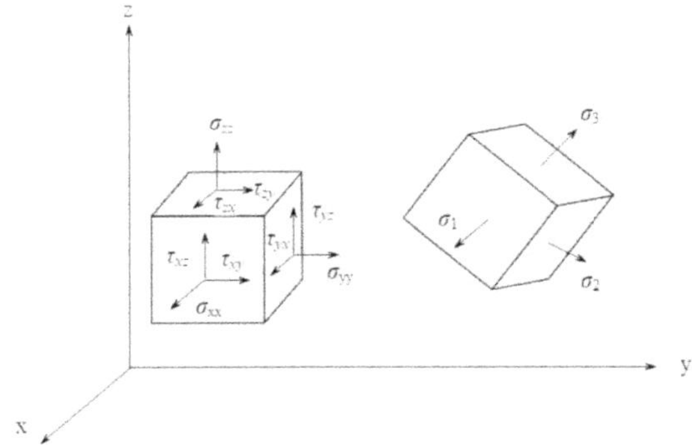

Figure 2.1 Coordinate stress and principal stress

Similarity to the principal stress, there are three invariants in the deviatoric stress tensor which are given by the following equations:

$$J_1 = s_{ii} = 0 \tag{2.27}$$

$$J_2 = \frac{1}{2}s_{ij}s_{ji} = \frac{1}{6}\left[(\sigma_{xx} - \sigma_{yy})^2 + (\sigma_{yy} - \sigma_{zz})^2 + (\sigma_{xx} - \sigma_{zz})^2\right] + \tau_{xy}^2 + \tau_{yz}^2 + \tau_{xz}^2 \tag{2.28}$$

$$J_3 = \det(s_{ij}) \tag{2.29}$$

The second invariant of deviatoric stress can be used for the calculation of the equivalent stress:

$$\sigma_v = \sqrt{3J_2} \tag{2.30}$$

The stress state for a point can be characterized by the principal stress and directions in the principal stress space as shown in Fig. 2.2.

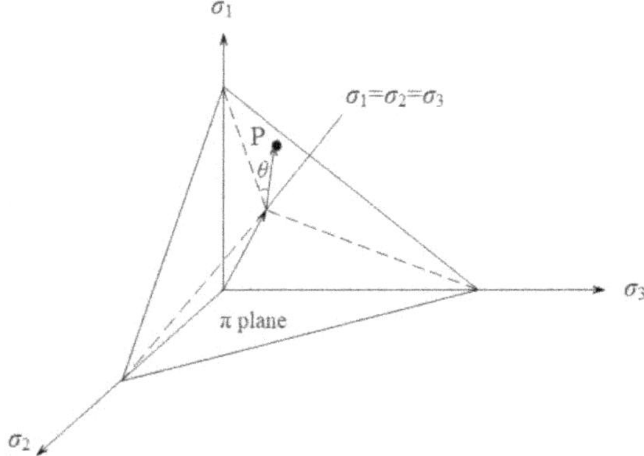

Figure 2.2 Principal stress space and stress geometry

The π-plane is the normal plane to the space diagonal. Obviously, the stress state on the space diagonal represents the isotropic stress state ($\sigma_1=\sigma_2=\sigma_3$). The stress state on the other occasions means the occurrence of deviatoric stress, which equals to the distance between the point and the diagonal. There are two special stress states including the triaxial compression ($\sigma_1\geq\sigma_2=\sigma_3$) and triaxial extension ($\sigma_1=\sigma_2\geq\sigma_3$). The other stress states locate between the states of triaxial compression and triaxial extension and can be characterize by the Lode number m and the Lode angle θ as follows:

$$m=\frac{2\sigma_2-(\sigma_1+\sigma_3)}{\sigma_1-\sigma_3}=\sqrt{3}\tan\theta \tag{2.31}$$

$$\sin 3\theta=-\frac{3\sqrt{3}J_3}{2\sqrt{J_2^3}} \tag{2.32}$$

2.2.2 Constitutive model

The relationship between the strain and the stress can be described by the constitutive equations. The total strain can be divided into elastic strain and plastic strain, which corresponding the part of strain before and after the stress reach to the yield point respectively. As can be seen in Fig. 2.3, there are four types of plastic stress-strain curves, including the strain-hardening, elastic-plastic, strain-softening, and brittle.

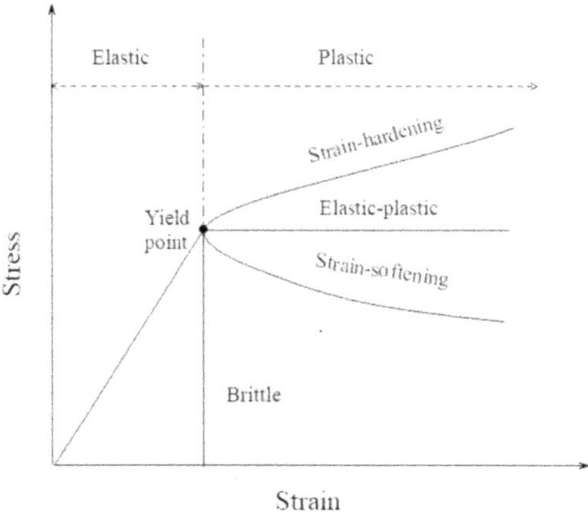

Figure 2.3 Typical plastic stress-strain curves

The elastic strain can be calculated by using the Hooke's low, which are given as follows:

$$\begin{cases} \varepsilon_{xx} = \frac{1}{E}\left[\sigma_{xx} - v\left(\sigma_{yy} + \sigma_{zz}\right)\right], & \varepsilon_{xy} = \frac{\tau_{xy}}{2G} \\ \varepsilon_{yy} = \frac{1}{E}\left[\sigma_{yy} - v\left(\sigma_{xx} + \sigma_{zz}\right)\right], & \varepsilon_{yz} = \frac{\tau_{yz}}{2G} \\ \varepsilon_{zz} = \frac{1}{E}\left[\sigma_{zz} - v\left(\sigma_{xx} + \sigma_{yy}\right)\right], & \varepsilon_{zx} = \frac{\tau_{zx}}{2G} \end{cases} \tag{2.33}$$

where E is the Young's modulus [Pa], v is the Poisson's ratio [-], G is the shear modulus [Pa].

To quantify the plastic strain, various of failure criterions such as the Mohr failure criterion, Drucker-Prager criterion and Hoek-Brown criterion have been proposed. The most widely used one is the Mohr failure criterion, in which the failure occurs when the shear stress exceed the shear strength. The expression of the Mohr failure criterion is given by:

$$F^P = \sigma_1 - \sigma_3 \cdot N_\varphi - 2c\sqrt{N_\varphi} \tag{2.34}$$

where σ_1 is the maximum principal stress [Pa], σ_3 is the minimum principal stress [Pa], c is the cohesion [Pa], N_φ is a constant and can be calculate as:

$$N_\varphi = \frac{1 + \sin\varphi}{1 - \sin\varphi} \tag{2.35}$$

where φ is the friction angle [°].

Based on the failure criterion, the plastic deformation can be obtained according to the following equation:

$$d\varepsilon_{ij}^{P} = d\lambda \cdot \frac{\langle F^P \rangle}{|F^P|} \cdot \frac{\partial Q^P}{\partial \sigma_{ij}} \tag{2.36}$$

$$\langle F^P \rangle = \begin{cases} 0 & F^P \le 0 \\ F^P & F^P > 0 \end{cases} \tag{2.37}$$

where λ is a plastic strain multiple factor [-], ε_{ij}^{P} is the plastic strain [-], Q^P is the potential function.

The potential function for the Mohr criterion is given as follows:

$$Q^P = \sigma_1 - \sigma_3 \cdot N_{\varphi} \tag{2.38}$$

2.3 THM coupling processes

The multiphase and multicomponent flow and the mechanical behavior of formation rock have been introduced independently. Yet, the processes of thermal, hydraulic, and mechanical (THM) would affect each other in the operation of geological CO_2 storage. The THM coupling concept will be presented in this section.

Fig. 2.4 shows the THM coupling concept in the operation of geological CO_2 storage. It can be divided into three parts, including the TH coupling, TM coupling, and HM coupling.

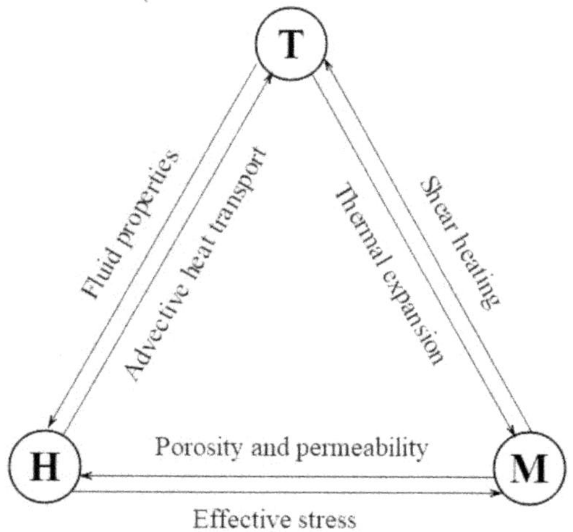

Figure 2.4 THM coupling concept

2.3.1 TH coupling model

The flowing of fluid in the reservoir would induce thermal effect and change the temperature, which can be characterized by the departure enthalpy and heat advection. Inversely, the heat transport would affect the hydraulic process through altering the fluid properties such as the density, viscosity, and the vapor pressure. There is a model based on the isobaric pore expansivity can be used to characterize the impact of thermal effect on the porosity, which is defined as the following equation:

$$\gamma_\phi = \frac{1}{\phi}\frac{\partial \phi}{\partial T} \tag{2.39}$$

Further, the impact on the permeability can be characterized by using the power-low relationship (Chin et al. 2000):

$$k = k_i\left(\frac{\phi}{\phi_i}\right)^n \tag{2.40}$$

where k_i represents the initial permeability [m^2], n denotes the material constant [-].

2.3.2 TM coupling model

The mechanical process would affect the thermal process by shear heating because of friction, while the effect is not significant. The effect of heat transport can also affect the mechanical process through the thermal stress caused by temperature change, which is given as follows:

$$\sigma^{th} = -3\beta K \Delta T \tag{2.41}$$

where K is the bulk modulus [Pa], β is the thermal expansion coefficient [1/°C], ΔT is the temperature change [°C].

2.3.3 HM coupling model

On the one hand, the hydraulic process would change the effective stress of formation rock by altering the fluid pressure. On the other hand, the mechanical process would affect the hydraulic process through altering the effective stress-dependent porosity and permeability. This is result from that the porosity, as well as the permeability can be compressed under a high effective stress.

The stress change by the hydraulic process can be calculated by:

$$\sigma^{hy} = \alpha P \tag{2.42}$$

where P is the formation fluid pressure [Pa], α is the Biot coefficient [-].

Substituting Eq. 2.41 and 2.42 into Eq. 2.33, the current pore pressure can be determined. Owing to the coupling effect, the new pressure can further cause stress redistribution in the formation rock, which can be described by the following equation:

$$\boldsymbol{\sigma} - (\alpha P + 3\beta K \Delta T)\mathbf{I} = \lambda tr\boldsymbol{\varepsilon}\mathbf{I} + 2\mu\boldsymbol{\varepsilon} \tag{2.43}$$

where σ denotes the current total stress tensor [Pa], $tr\varepsilon$ represents the trace of the strain tensor ε [-], λ is the Lamé parameter [Pa], **I** represents the unit tensor, and μ denotes the shear modulus [Pa].

According to the work conducted by Rutqvist and Tsang (2002), the stress-dependent porosity and the permeability can be expressed by Eq. 2.44 and 2.45, respectively. Actually, these mathematical models are modified from the original work conducted by Davies and Davies (1999) and derived based on the data for 1000 direct laboratory measurements of permeability in core plug samples for various conditions of net effective stress.

$$\phi = (\phi_0 - \phi_r)\exp\left(5\times10^{-8}\cdot\sigma_M'\right) + \phi_r \tag{2.44}$$

$$k = k_0 \exp\left[22.2\left(\frac{\phi}{\phi_0} - 1\right)\right] \tag{2.45}$$

where ϕ_0 denotes the porosity at the zero-stress condition [-], ϕ_r represents the residual porosity [-], $\sigma_M' = \sigma_M - \alpha P$ is the effective mean stress [Pa], σ_M denotes the mean stress [Pa], and k_0 represents the permeability at the zero-stress condition [m^2].

2.4 THM coupling models in TOUGH2MP (TMVOC)-FLAC3D

2.4.1 TOUGH2MP (TMVOC)

► TOUGH2MP

TOUGH2MP is a massively parallel version of TOUGH2 code, which is developed by the Lawrence Berkeley National Laboratory and enables the simulation of multi-component and multi-phase flow in both isothermal and non-isothermal conditions (Zhang et al. 2008). TOUGH2MP has a good performance on solving the large or highly nonlinear problems, which is suitable for the simulation of geological CO_2 storage.

In TOUGH2MP, the integral finite difference method was used to discretize the continuum equations in space. The appropriate volume average was introduced as follows (Pruess et al. 1999):

$$\int_{V_n} M dV = V_n M_n \tag{2.46}$$

where V_n denotes the control volume for an arbitrary subdomain in the flow system [m³], M represents the volume-normalized extensive quantity [m³], M_n denotes the average value of M over the V_n [m³].

The surface integrals are obtained by the sum of averages over the surface segments A_{nm}:

$$\int_{\Gamma_n} \mathbf{F}^K \cdot \mathbf{n} d\Gamma = \sum_m A_{nm} F_{nm} \tag{2.47}$$

where F_{nm} denotes the average value of normal component over the surface between the two volume elements, i.e., V_n and V_m. Fig. 2.5 shows the space discretization and data in the integral finite difference method.

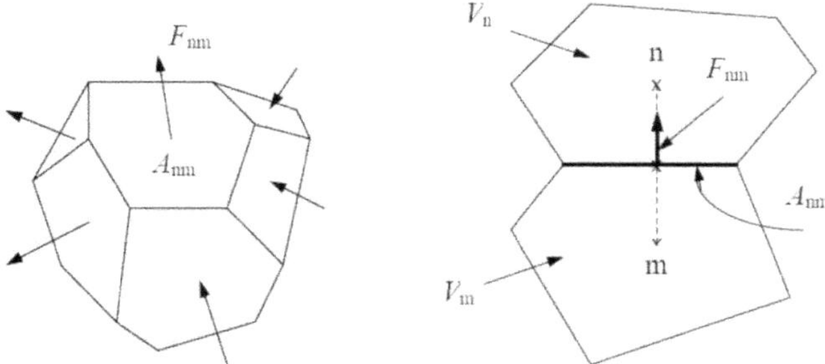

Figure 2.5 Schematic of the integral finite difference method in the TOUGH2MP (Pruess et al. 1999)

Substituting Eqs. 2.44 and 2.45 into the governing equation (Eq. 2.1), the ordinary differential equations in time can be obtained according to the following equation:

$$\frac{dM_n^k}{dt} = \frac{1}{V_n} \sum_m A_{nm} F_{nm}^k + q_n^k \tag{2.48}$$

To calculate the equations efficiently, the sink and source term q_n^k is evaluated at a new time level of

$$t^{k+1} = t^k + \Delta t \tag{2.49}$$

The mass and energy conservation equations are solved by using the Newton iteration. For component k, its residuum term $R_n^{\kappa,t+1}$ at the time of t+1 is obtained on the basis of the change in the accumulation terms for a time step Δt. The flux and injection source at the time of t+1 can be calculated as follows:

$$R_n^{k,t+1} = M_n^{k,t+1} - M_n^{k,t} - \frac{\Delta t}{V_n} \sum_m A_{mn} F_{mn}^{k,t+1} + \Delta t q_n^{k,t+1} = 0 \tag{2.50}$$

After the calculation of each iteration step, all the secondary parameters including the density, capillary pressure, and the relative permeability are updated according to the equation of state (EOS) obtained from the TMVOC EOS (Pruess and Battistelli 2002).

The total multiphase diffusive flux (Eq. 2.14) can be discretized as:

$$\left(f^k\right)_{mn} = -\left(\Sigma_1^k\right)_{nm} \frac{\left(X_1^k\right)_m - \left(X_1^k\right)_n}{D_{nm}} - \left(\Sigma_g^k\right)_{nm} \frac{\left(X_g^k\right)_m - \left(X_g^k\right)_n}{D_{nm}} \tag{2.51}$$

Eq. 2.51 can be modified as a single mass fraction gradient and an effective diffusive strength coefficient. For example, the liquid mass fraction can be expressed by the following equation:

$$\left(f^k\right)_{nm} = -\left\{\Sigma_1^k + \Sigma_g^K \frac{\left(X_g^k\right)_m - \left(X_g^k\right)_n}{\left(X_1^k\right)_m - \left(X_1^k\right)_n}\right\}_{nm} \frac{\left(X_1^k\right)_m - \left(X_1^k\right)_n}{D_{nm}} \tag{2.52}$$

► **TMVOC**

TMVOC is an extension of the TOUGH program developed by the Lawrence Berkeley National Laboratory (Pruess and Battistelli 2002). In the formulation of TMVOC, the multiphase system is composed of water, volatile organic chemicals, and non-condensable gases. Regarding the non-condensable gases in TMVOC, CO_2, CH_4, O_2, and N_2 are included. Therefore, it can be used for the simulation of the geological CO_2 storage. The modular architecture of TMVOC is shown in Fig. 2.6.

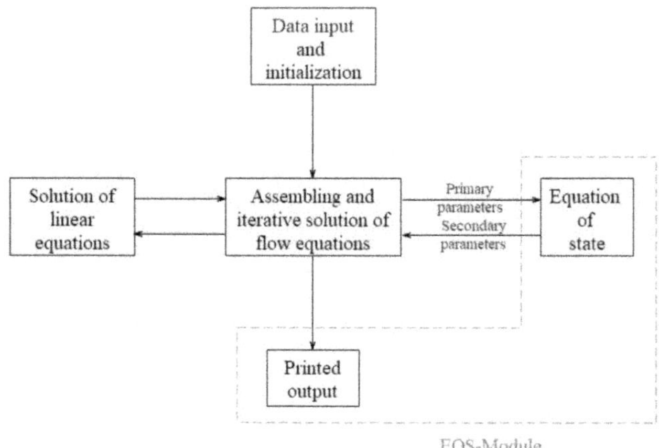

Figure 2.6 The modular architecture of TMVOC (Pruess and Battistelli 2002)

Firstly, the primary thermodynamic parameters such as temperature, pressure, and saturation are assembled for all the grid blocks. Meanwhile, the secondary thermodynamic parameters such as the density and solubility are distributed for the flow and transport equations. After solving the flow equations in a time step, the updated primary parameters will be sent to the equations of state to get updated secondary parameters, which will be sent to the flow equations for calculation again. The execution of the program will continue until the termination criteria is reached, such as a specified simulation time or a total number of time steps.

► **Verification of TOUGH2MP (TMVOC)**

- **Generic model from a literature study**

To validate the performance of TOUGH2MP (TMVOC) on simulating the underground flowing of CO_2 and the mixing behavior of CO_2 and CH_4, the case and simulation results from Oldenburg (2003) and Ma et al. (2019) are used and compared. Fig. 2.7 shows the sketch of the reservoir model and boundary conditions. The geological model consists of a two-dimensional reservoir with a length of 1000 m and a thickness of 22 m. The initial pressure of the reservoir is 6 MPa, and the reservoir temperature is 40 °C. The reservoir is initially saturated with CO_2. All the boundaries of the model are set as no flow boundary. The model is discretized into 4400 (200 × 22) grid blocks. Only a single well is applied for both injection and production well. The CH_4 is injected at a constant rate of 1.8375×10^{-2} kg/s for 180 days.

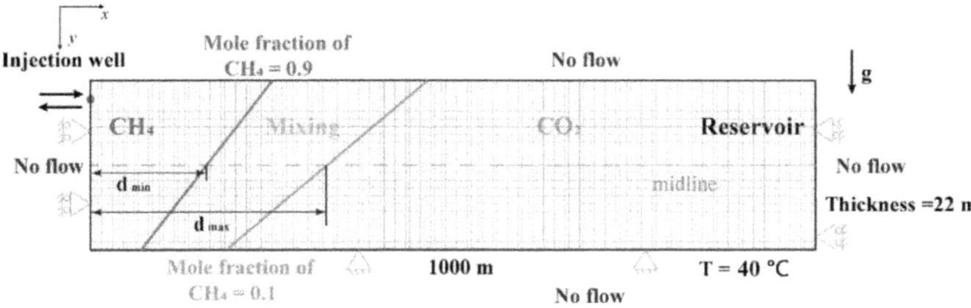

Figure 2.7 Sketch of the reservoir model and boundary conditions (Ma et al. 2019)

- **Simulation results in comparison with the results from the literature study**

As can be seen in Fig. 2.7, the mixing zone is regarded as the region with a CH_4 mole fraction ranges from 0.1 to 0.9. The d_{min} and d_{max} denote the distance between the left boundary with the nearest and farthest extents of the mixing region at the depth of the reference line. The simulation results from TOUGH2MP (TMVOC) are compared with Oldenburg (2003)'s and Ma et al. (2019)'s results in Tab. 2.1. It can be seen that the d_{min} and d_{max} are 108.0 m and 226.8 m after 30 days of injection, 547.71 m and 750.07 m after 180 days of injection. The average migration distance is 167.4 m and 648.9 m for 30 and 180 days of injection, respectively. The relative error between these data and the results of Oldenburg (2003) and Ma et al. (2019) only ranges from 1.05% to 11.57%, which demonstrates that the TOUGH2MP (TMVOC) can be used to characterize the underground flowing of CO_2 and the mixing behavior of CO_2 and CH_4.

Table 2.1 Comparison of the simulation results

	30d			180d		
	d_{min}	d_{max}	d_{ave}	d_{min}	d_{max}	d_{ave}
Present	108.0	226.8	167.4	547.71	750.07	648.9
Ma et al. 2019	116.9	234.2	175.55	530.3	786.8	658.55
Relative error	7.62%	3.16%	4.64%	-3.28%	4.67%	1.47%
Oldenburg 2003	104.63	246.64	175.64	484.35	742.18	613.26
Relative error	-3.12%	8.75%	4.92%	-11.57%	-1.05%	-5.49%

2.4.2 FLAC3D

FLAC3D is a three-dimensional program for engineering mechanics computational simulation based on the finite difference method, which is developed by the Itasca Consulting Group, Inc. (Itasca 2009). FLAC3D enables the linear or nonlinear stress-strain relationship in response to the restraints of boundaries and the applied forces. The explicit Lagrangian calculation scheme and the mixed-discretization zoning technique including the finite difference approach, the discrete-model approach, and the dynamic-solution approach were used in FLAC3D (Cundall 2008). In FLAC3D, the element will be transformed into sub tetrahedron element as shown in Fig. 2.8.

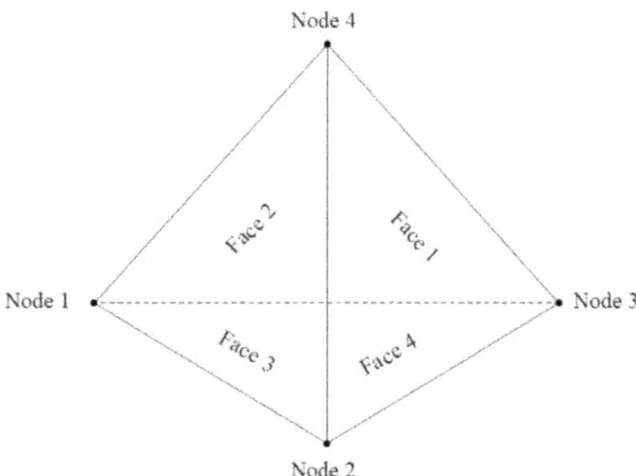

Figure 2.8 Transformed tetrahedron in FLAC3D

The volume integrals in the tetrahedron can be expressed by the integrals of the surface of tetrahedron according to the Gauss divergence theorem:

$$\int_V v_{i,j} dV = \int_S v_i n_j dS \tag{2.53}$$

where n_j denotes the exterior unit vector normal to the corresponding surface.

After integration, Eq. 2.53 yields the following equations:

$$Vv_{i,j} = \sum_{f=1}^{4} \overline{v}_i^{(f)} n_j^{(f)} S^{(f)} \tag{2.54}$$

where $\overline{v_i}$ represents the average value of velocity, f denotes the variable on face f.

For a linear velocity variation, Eq. 2.54 can be converted into:

$$Vv_{i,j} = \frac{1}{3}\sum_{l=1}^{4} v_i^l \sum_{f=1, f\neq l}^{4} n_j^{(f)} S^{(f)} \tag{2.55}$$

By using the divergence theorem, the velocity can be expressed as:

$$v_{i,j} = -\frac{1}{3V}\sum_{l=1}^{4} v_i^l n_j^{(l)} S^{(l)} \tag{2.56}$$

For an integral domain, we can write:

$$\int_{\Omega} \boldsymbol{\sigma} : \Delta\dot{\varepsilon} d\Omega = \int_{\Gamma_f} f \cdot \Delta v d\Gamma_f + \int_{\Omega} \rho\left(b - \frac{dv}{dt}\right) \Delta v d\Omega \tag{2.57}$$

where $\boldsymbol{\sigma} : \Delta\dot{\varepsilon}$ denotes the internal work rate, $\rho\left(b - \frac{dv}{dt}\right)\Delta v$ represents the external work rate due to velocity change, $f \cdot \Delta v$ is the external force work rate, Γ_f denotes the integral boundary of force, $\dot{\varepsilon}$ is the strain increment rate tensor and can be written as the following equations:

$$\Delta\dot{\varepsilon}_{ij} = \frac{1}{2}(v_{i,j} + v_{j,i}) \tag{2.58}$$

The stress increment tensor can be expressed as:

$$\Delta\boldsymbol{\sigma} = \mathbf{D}\Delta\boldsymbol{\varepsilon} \tag{2.59}$$

where **D** is the physical matrix.

Replace the strain increment rate by Eq. 2.58 in the Eq. 2.56 and introduce the velocity divergence, the internal work rate in a tetrahedron element can be written as the following equation:

$$\int_{\Omega} \boldsymbol{\sigma} : \Delta\dot{\varepsilon} d\Omega = -\frac{1}{3}\sum_{l=1}^{4} \Delta v_i^l T_i^l \tag{2.60}$$

where $T_i^l = \sigma_{ij} n_j^{(l)} S^{(l)}$

The work rate for the external force at the node can be expressed as:

$$\int_{\Gamma_f} f_i \Delta v_i d\Gamma_f = \sum_{l=1}^{4} \Delta v_i^l f_i^l \tag{2.61}$$

Likewise, the contribution of work rate for the body farce on a tetrahedron element is given as:

$$\int_{\Omega} \Delta v_i \rho b_i d\Omega = \sum_{l=1}^{4} \Delta v_i^l m^l b_i \tag{2.62}$$

where $m^l = \frac{\rho V}{4}$, v denotes the volume of the tetrahedron element.

The contribution of work rate for velocity change can be written as:

$$\int_{\Omega} \Delta v_i \rho \frac{dv}{dt} d\Omega = \sum_{l=1}^{4} \Delta v_i^l m^l \left(\frac{dv_i}{dt} \right)^l \tag{2.63}$$

Substituting Eqs. 2.58-2.61 to Eq. 2.53, the work equilibrium for the node l can be expressed as the following general form:

$$m^l \left(\frac{dv_i}{dt} \right)^l = \frac{T_i^l}{3} + m^l b_i + f_i^l = F_i^l \tag{2.64}$$

where F_i^l denotes the total global contribution for node l.

The disturbance force would change the movement state of the grid point until the static state or quasi-static state is reached caused by the damping force, whose direction is opposite to the movement. The damping force can be expressed as follows:

$$Fd_i^l = -\alpha \left| F_i^l \right| \mathrm{sign}(v_i^l) \tag{2.65}$$

$$\mathrm{sign}(x) = \begin{cases} 1 & x > 0 \\ -1 & x < 0 \\ 0 & x = 0 \end{cases} \tag{2.66}$$

Thus, the condition of the static state or quasi-static state can be written as:

$$m^l \left(\frac{dv_i}{dt} \right)^l = F_i^l + Fd_i^l \tag{2.67}$$

The velocity of node l at the next time step can be obtained by the calculation of the following equation:

$$v_i^l(t + \Delta t) = v_i^l(t) + \frac{\Delta t}{m^l} \left(F_i^l(t) + Fd_i^l(t) \right) \tag{2.68}$$

Based on the time step and new node velocity, the displacement increment of the node at the time of $t+\Delta t$ can be obtained. Subsequently, the displacement increment can be used to calculate the new strain and stress increment according to Eqs. 2.56 and 2.57.

2.4.3 Coupling of TOUGH2MP (TMVOC) and FLAC3D

The THM coupling of TOUGH2MP-FLAC3D is developed based on the previous studies (Gou et al. 2014; Rutqvist and Tsang 2003, 2005). Fig. 2.9 shows the flowchart of the THM coupled simulator TOUGH2MP(TMVOC)-FLAC3D. After the governing equations have been solved (Eq. 2.1-2.10), the primary variables such as the temperature, pressure, and liquid saturation are obtained from the sub-processors of the parallel code TOUGH2MP firstly. Then they will be sent to FLAC3D to update the stress and strain (Eq. 2.43) in the formation rock for every time step. Further, the new permeability and porosity affected by the effective stress (Eq. 2.44 and 2.45) will be calculated and sent to TOUGH2MP. The data received in TOUGH2MP will be allocated to each sub-processor to calculate the corresponding secondary parameters, i.e., density, viscosity, and solubility, based on the equation of state in the TMVOC module. Under the basis of these secondary parameters and the injection rate, the new primary variables at the next time step can be obtained. The execution of the program will continue until the time limit is reached.

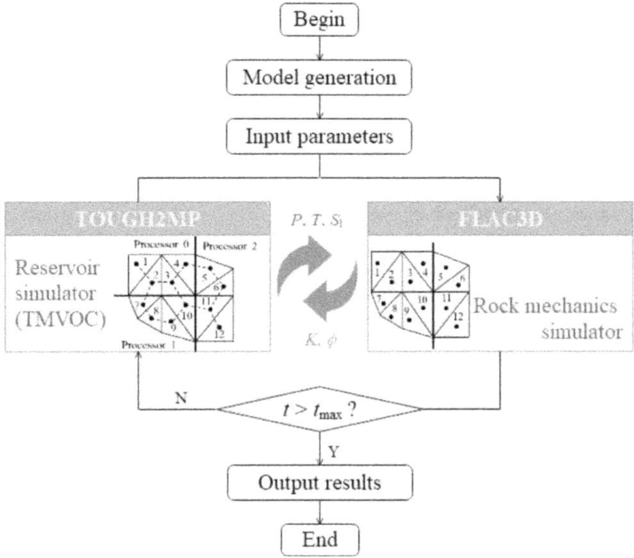

Figure 2.9 Flowchart of the TOUGH2MP (TMVOC)-FLAC3D simulator (Cao et al. 2020)

3 Parametric uncertainty analysis for CO_2 storage

In this chapter, a generic 2D model with geometric parameters that are averaged from worldwide large-scale CCS projects is used. To quantitatively analyze the role of uncertainty parameters within a possible range during CO_2 injection, the geomechanical and hydrogeological parameters, including Young's modulus, Poisson's ratio, Biot coefficient, permeability, permeability anisotropy ratio, and the injection rate were compiled from the literature and randomly sampled for 238 data points by using the Quasi-Monte Carlo method. Numerical simulation was performed on every data point using the developed TOUGH2MP(TMVOC)-FLAC3D simulator. The correlation between the simulated results and different parameters was analyzed based on distance correlation (Székely et al. 2007) to determine the key parameters for different formation responses, i.e., the formation fluid pressure and formation deformation, as well as the risk factors. In addition, a support vector regression (SVR) surrogate model was developed based on the machine learning approach in SVR (Chang and Lin 2011; Drucker et al. 1997; Vapnik 2013) and verified by using the numerical simulation results. Thereafter, the impact on the formation response, including pressure change and deformation, was predicted by using the trained SVR surrogate model for different values of the key parameters. The main contents of this chapter have been published in the following research paper (Cao et al. 2020): Parametric uncertainty analysis for CO_2 sequestration based on distance correlation and support vector regression. *Journal of Natural Gas Science and Engineering*, 103237.

3.1 Methodology

3.1.1 Distance correlation

Distance correlation was introduced by Székely et al. (2007) and can be used to characterize the dependence between random vectors. It is used to measure the dependence between uncertainties and formation responses. In comparison with the Pearson product-moment correlation coefficient, the distance correlation has a significant benefit for characterizing a non-linear relationship. According to previous research (Li et al. 2018; Newell et al. 2017; Rutqvist et al. 2010), a non-linear relationship is usually determined between the reservoir parameters and formation responses, including the pressure change and deformation during CO_2 injection. Therefore, distance correlation could be an excellent indicator of corresponding key parameters in the case of non-linear correlations. To the best of my knowledge, this is the first time for the distance correlation to be used in the parametric uncertainties analysis for CO_2 sequestration.

The definition of the empirical distance correlation based on the work conducted by Székely et al. (2007) is given in the following. Set one of the sampled parameter datasets as vector **X**, and another corresponding dataset as vector **Y**. Using the variables in vector **X**, then the pairwise distance (α_{kl}), row mean of k-th ($\overline{a}_{k.}$), column mean of l-th ($\overline{a}_{.l}$), as well as the grand mean ($\overline{a}_{..}$) could be calculated separately by using Eqs. 3.1-3.4. Based on these data, the centered distance (A_{kl}) of vector **X** can be obtained by solving Eq. 3.5.

$$a_{kl} = |X_k - X_l| \tag{3.1}$$

$$\bar{a}_{k.} = \frac{1}{n}\sum_{l=1}^{n} a_{kl} \tag{3.2}$$

$$\bar{a}_{.l} = \frac{1}{n}\sum_{k=1}^{n} a_{kl} \tag{3.3}$$

$$\bar{a}_{..} = \frac{1}{n^2}\sum_{k,l=1}^{n} a_{kl} \tag{3.4}$$

$$A_{kl} = a_{kl} - \bar{a}_{k.} - \bar{a}_{.l} + \bar{a}_{..} \tag{3.5}$$

and $k,l = 1,\ldots,n$, where n represents the number of the element in **X**.

Similarly, for the vector **Y**, the corresponding parameters $b_{kl} = |Y_k - Y_l|$ and $B_{kl} = b_{kl} - \bar{b}_{k.} - \bar{b}_{.l} + \bar{b}$ can be obtained by using Eqs. 3.1-3.4. The empirical distance covariance $v_n(\mathbf{X},\mathbf{Y})$ is then given as the following equation:

$$\mathcal{V}_n^2(\mathbf{X},\mathbf{Y}) = \frac{1}{n^2}\sum_{k,l=1}^{n} A_{kl}B_{kl} \tag{3.6}$$

Correspondingly, the empirical distance variance of vector X and Y can be calculated as follows:

$$\mathcal{V}_n^2(\mathbf{X}) = \mathcal{V}_n^2(\mathbf{X},\mathbf{X}) = \frac{1}{n^2}\sum_{k,l=1}^{n} A_{kl}^2 \tag{3.7}$$

$$\mathcal{V}_n^2(\mathbf{Y}) = \mathcal{V}_n^2(\mathbf{Y},\mathbf{Y}) = \frac{1}{n^2}\sum_{k,l=1}^{n} B_{kl}^2 \tag{3.8}$$

The empirical distance correlation $\mathcal{R}_n(\mathbf{X},\mathbf{Y})$ between vector **X** and **Y** can be obtained according to Eq. 3.9.

$$\mathcal{R}_n(\mathbf{X},\mathbf{Y}) = \begin{cases} \sqrt{\dfrac{\mathcal{V}_n^2(\mathbf{X},\mathbf{Y})}{\sqrt{\mathcal{V}_n^2(\mathbf{X})\mathcal{V}_n^2(\mathbf{Y})}}}, & \mathbf{V}_n^2(\mathbf{X})\mathcal{V}_n^2(\mathbf{Y}) > 0 \\ 0, & \mathbf{V}_n^2(\mathbf{X})\mathcal{V}_n^2(\mathbf{Y}) = 0 \end{cases} \tag{3.9}$$

This distance correlation can be used to measure the dependence between vector **X** and **Y**. The dependence increases with the value of distance correlation. If the distance correlation is equal to 1, this implies that there is a vector **A**, scalar *b*, and orthonormal matrix **C** such that **Y**=**A**+*b***CX**. The distance correlation coefficient is zero, if and only if these random vectors are independent. However, if the value of distance correlation approaches 1, this indicates that there is a correlation between the vectors. In this thesis, the geomechanical and hydrogeological parameters are set as vector **X**, and the pressure change and deformation are set as vector **Y**. Therefore, the corresponding correlation can be obtained by calculating the distance correlation.

3.1.2 Support vector regression

Due to the parameters considered may vary in a certain range, if a total of ten parameters including the Young's modulus, Poisson's ratio, Biot coefficient, permeability and anisotropy ratio of the reservoir and caprock, and injection rate, varies simultaneously, there will be a lot of simulation work in the originally described physical model. Specifically, the physical model needs to be simulated for 3^{10} times if 3 value of each parameter is considered. Even though statistical methods are employed to design the numerical experiment, it would be a lot of simulation work and cost a lot of time (several days). To address this issue, the surrogate models based on SVR are developed to acquire the formation response, i.e., formation pressure and formation displacement, to CO_2 injection. By using the surrogate models, the results can be obtained within several seconds.

Due to its high performance in solving nonlinear regression problems, the machine learning method has been widely used in SVR (Chang and Lin 2011; Drucker et al. 1997; Vapnik 2013). The critical idea of SVR is the mapping of input patterns into a higher dimensional feature space by using a nonlinear mapping approach. A linear regression model in the corresponding feature space then can be obtained (Burges 1998). The typical process of SVR consists of two steps. The first step involves providing a training dataset that contains training input data and output data to obtain a trained model. The second step is to use the trained model to predict the output response of a given input dataset for testing. The theory of SVR is briefly presented as follows (Chang and Lin 2011).

For a set of training datasets, $\{(\mathbf{x}_1, y_1),\ldots,(\mathbf{x}_l, y_l)\}$ from the true model, where $\mathbf{x}_i \in R^n$ is the feature vector, and $y_i \in R^1$ is the output data. The standard form of the SVR under the given parameters $C > 0$ and $\epsilon > 0$ can be described as follows (Chang and Lin 2011; Vapnik 2013):

$$\min_{w,b,\xi,\xi^*} \frac{1}{2} w^T w + C\sum_{i=1}^{l} \xi_i + C\sum_{i=1}^{l} \xi_i^* \tag{3.10}$$

Subject to the constraints outlined as follows:

$$\begin{cases} w^T\phi(x_i)+b-y_i \le \epsilon+\xi_i \\ y_i - w^T\phi(x_i)-b \le \epsilon+\xi_i^* \\ \xi_i, \xi_i^* \ge 0, i=1,\dots,l \end{cases} \tag{3.11}$$

where w represents the vector variable, C is the trade-off between regularization and the tube violation, ξ_i represents the upper tube violations, ξ_i^* denotes the lower tube violations, b denotes the parameter that determines the linear parameterization, ϵ represents the vertical tube width, and $\phi(x_i)$ maps x_i into a higher-dimensional space.

Due to the possible high dimensionality of the vector variable w in the standard form, it would be difficult to be solved. To address this issue, the dual problem that can be obtained by using the optimization method is proposed, which is easily solved compared with the primal problem. The expression of the dual problem can be expressed as:

$$\min_{\alpha,\alpha^*} \frac{1}{2}(\alpha-\alpha^*)^T Q(\alpha-\alpha^*)+\epsilon\sum_{i=1}^{l}(\alpha_i+\alpha_i^*)+\sum_{i=1}^{l} y_i(\alpha_i-\alpha_i^*) \tag{3.12}$$

with the constraints:

$$\begin{cases} \mathbf{e}^T(\boldsymbol{\alpha}-\boldsymbol{\alpha}^*)=0 \\ 0 \le \alpha_i, \alpha_i^* \le C, i=1,\dots,l \end{cases} \tag{3.13}$$

where α and α^* are the dual variables for each data point. $Q_{ij}=K(\mathbf{x}_i,\mathbf{x}_j)\equiv\phi(\mathbf{x}_i)^T\phi(\mathbf{x}_j)$. $\mathbf{e}$ denotes a vector of all ones.

After the dual problem is solved, the SVR function can be written as:

$$f(x)=\sum_{i=1}^{l}(-\alpha_i+\alpha_i^*)K(\mathbf{x}_i,\mathbf{x})+b \tag{3.14}$$

where $K(\mathbf{x}_i, \mathbf{x})$ represents the Kernel function. In this thesis, the Sigmoid kernel was used as the Kernel function, which can be written as

$$K(\mathbf{x}_i,\mathbf{x})=\tanh(a\mathbf{x}_i^T\mathbf{x}+r) \tag{3.15}$$

where a represents the scaling parameter of the input data, and r denotes the shifting parameter that controls the threshold of mapping. In addition to the Kernel function parameters, another important parameter in the training of SVR is the penalty factor, which means the tolerance to the error. It should be mentioned that both higher and lower value of penalty factor would lead to worse generalization, which is harmful to the training of the surrogate model. One the one hand, the error is more likely not to be tolerated with a higher value of penalty factor, which probably resulting overfitting. On the other hand, lower value of penalty factor may lead to underfitting. Therefore, the value of penalty factor is supposed to be optimized. In this thesis, the optimization of the penalty factor and Kernel function parameter were conducted through the method of cross-validation.

In this study, the geomechanical and hydrogeological parameters are set as the feature vectors, whereas the pressure change and formation deformation are set as the output data. Therefore, the SVR surrogate model can be obtained based on the SVR. Then the SVR surrogate model can be used to predict the pressure change and formation deformation.

3.1.3 Working Schema

To determine the impact of different geomechanical and hydrogeological parameters on the reservoir response, this work is divided into three parts according to the different methods utilized, including simulation, distance correlation analysis, and SVR. As shown in Fig. 3.1, important parameters with typical ranges are initially collected to perform simulations based on a literature study. The collected parameters are then randomly sampled into many sample points. Based on this, simulations are conducted by using the THM coupled simulator to acquire corresponding results. For the distance correlation analysis, the sample points and the simulated results are set as the input dataset to identify the parameters that affect the reservoir response. Moreover, the SVR surrogate models are trained on the same input dataset of the normalized sample points and the output dataset of the simulated results by using SVR. The trained SVR surrogate model is used to predict the reservoir response for much denser sample points, in which only key parameters are sampled and other parameters are kept constant to elucidate the role of key parameters.

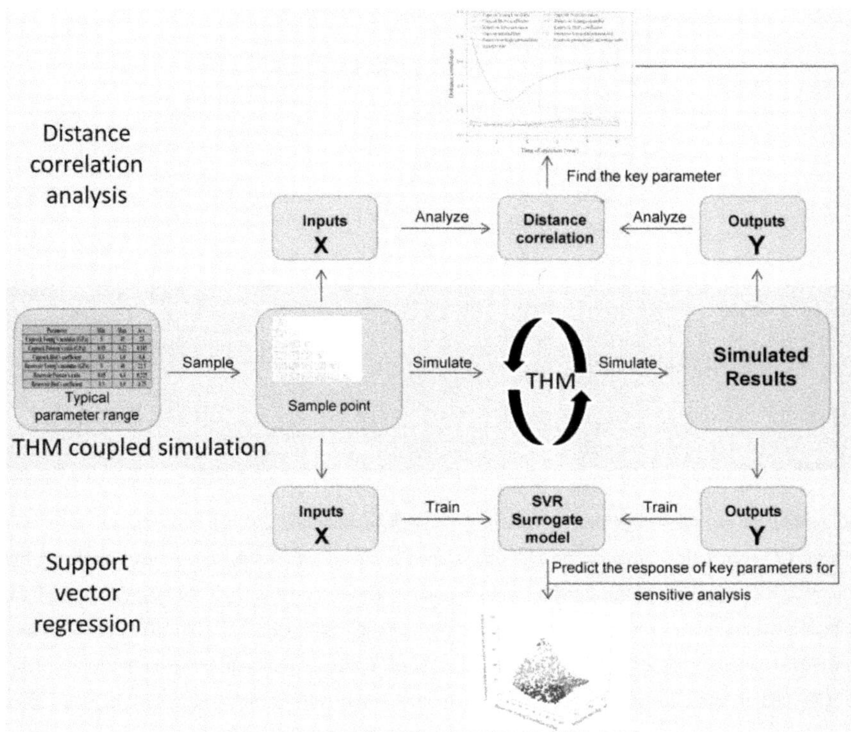

Figure 3.1 Overview of the different aspects of the proposed approach (Cao et al. 2020)

3.2 Simulation schemes

3.2.1 Generic model based on worldwide CCS projects

There are some large-scale CCS projects been operated and planned worldwide, including the Snøhvit project (Hansen et al. 2013), Sleipner project (Audigane et al. 2006, 2007; Williams and Chadwick 2018), Salah project (Ringrose et al. 2013; Rutqvist et al. 2010), Gorgon project (Flett et al. 2008), and Quest project (Bourne et al. 2014). The main geological and hydrogeological parameters of these CCS projects are summarized in Tab. 3.1. A representative generic model for CCS, with a depth of 1930 m, thickness of 130 m, caprock thickness of 270 m, and temperature of 75 °C, was obtained based on an average of the forementioned five CCS projects.

Table 3.1 Main large-scale parameters used for worldwide CCS projects (Cao et al. 2020)

Num.	Project	Injection rate[t/d]	K [mD]	Depth [m]	Thickness [m]	Thickness [m]	T [°C]	P [MPa]	Source
1	Snøhvit	2000	450	2550	60	30	95	28.5	Hansen et al. (2013); Michael et al. (2010)
2	Sleipner	2700	3000	1000	250	75	37	10.3	Arts et al. 2008; Audigane et al. 2006; Michael et al. (2010
3	In Salah	3500	13	1800	20	900	90	17.9	Michael et al. 2010; Rutqvist et al. 2010
4	Gorgon	10410	25	2300	280	250	100	22	Flett et al. 2008; Michael et al. 2010
5	Quest	2960	100	2000	40	70	55	18.9	Bachu 2013; Bourne et al. 2014; Huang et al. 2014
Ave. or Range	/	2000 ~ 10410	13 ~ 3000	1930	130	270	75	19.5	/

Regarding the boundary conditions, there are two boundary conditions that can be considered in the two phases CO_2-water flow system. The first one is open boundary condition, which means allowing the fluids flow into and out of the domain closed boundary. The second one is closed boundary condition, which means no mass or heat flux is allowed to pass through the boundaries (Liu 2014). In this work, the closed boundary is used in the model. For the purpose of providing enough space for the transmission of CO_2 and pressure, the infinite boundary was applied in this work, which is also used in many other research works (Bao et al. 2013; Rutqvist and Tsang 2002). This generic model has a length up to 40,000 m, which was used to build an infinite boundary condition. To minimize the computational burden, a 2D model (40,000 m × 3500 m) with a thickness of 100 m was generated as shown in Fig. 3.2.

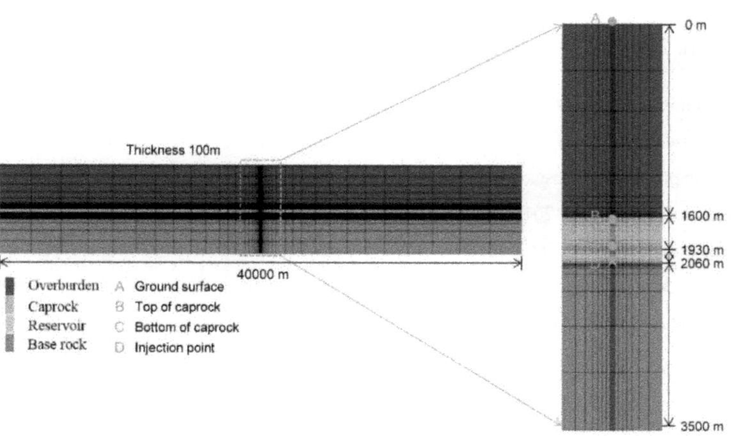

Figure 3.2 Generic model for simulation and the key points for response measurements (Cao et al. 2020)

In the CCS systems, the security of CO_2 storage is dependent on the integrity of the caprock if the wellbore is integrated and stable (Kreft et al. 2007). In other words, as long as the integrity of caprock was guaranteed, the injected CO_2 cannot migrate upward. According to the work of Gou et al. (2016), the region with maximum pressure increment in the caprock located at the bottom zone as can be seen in Fig. 3.3, which means that the point at the bottom of caprock is more likely to break the caprock integrity. This is also identified by the maximum fluid pressure in caprock located at the bottom zone. As expected, Fig. 3.3 also shows that the reservoir (injection zone) has a high fluid pressure. Therefore, the bottom of caprock (C in Fig. 3.3) and injection zone (D) can be used for representativeness of the fluid pressure for caprock and reservoir model respectively. Regarding the formation deformation, the maximum vertical displacement is located at the above of the injection zone (Rutqvist et al. 2010). In addition, the displacement at the ground surface is another important indicator for the formation deformation (Rinaldi et al. 2017), which can be monitored by the satellite-based inferrometry such as the application in the In Salah Gas Project (Rutqvist et al. 2010). Therefore, the top (B in Fig. 3.3) and bottom (C) of caprock and ground surface (A) are identified as key points for the detection of the formation deformation, i.e., vertical displacement to CO_2 injection.

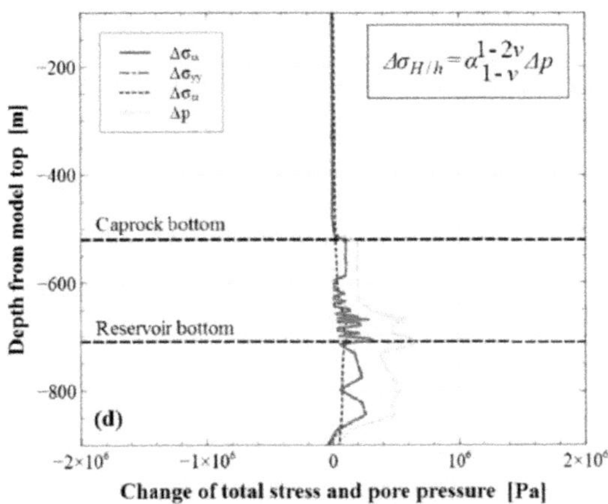

Figure 3.3 Stress change caused by CO2 injection along the vertical line through the injection zone (Gou et al. 2016)

In this model, a no-flow boundary was utilized, and the displacement of the four sides and the bottom boundary was fixed at zero. The initial pore pressure was converted from the hydrostatic pressure, and the formation pores were initially saturated with water. It should be mentioned that the CO_2 is in the supercritical state under the reservoir temperature and pressure conditions. As shown in Fig. 3.4, the CO_2 shows a low viscosity of approximately 0.5×10^{-4} Pa·s and high density of approximately 700 kg/m³ under the mentioned temperature and pressure conditions.

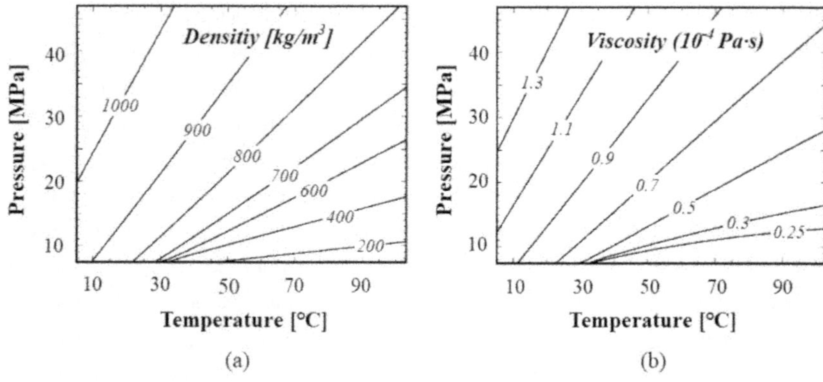

Figure 3.4 Contour diagram of (a) CO_2 density and (b) viscosity (Pruess and García 2002)

In the two phases CO_2-water flow system, the modified SRK cubic equation of state (Soave 1972) is adopted to calculate the parameters of the gases, which can be expressed as follows:

$$p = \frac{RT}{v-b} - \frac{a\alpha(T)}{v(v+b)} \tag{3.16}$$

where p is the pressure, T denotes the temperature, R is the universal gas constant, v represents the volume. a and b are the parameters dependent on the critical temperature T_c, the critical pressure p_c that can be calculated according to the following equations:

$$a = 0.42747\frac{R^2T_c^2}{p_c} \tag{3.17}$$

$$b = 0.08664\frac{RT_c}{p_c} \tag{3.18}$$

The expression of $\alpha(T)$ can be written as:

$$\alpha(T) = \left(1 + (0.48 + 1.574\omega - 0.176\omega^2)(1 - \sqrt{\frac{T}{T_c}})\right)^2 \tag{3.19}$$

where ω is the acentric factor.

The relative permeability of a gas and liquid are calculated according to Corey's model (Corey 1954), and van Genuchten's function (Van Genuchten 1980) in Eqs. 2.7 and 2.8, respectively. A poro-elastic homogeneous medium was used for all the formations. Moreover, the simulation was conducted under an isothermal condition and an isotropic stress regime with a gradient of 22.17 MPa/km.

3.2.2 Simulation properties

The geomechanical parameters (Young's modulus, Poisson's ratio, and Biot coefficient) and the hydrogeological parameters (permeability and its anisotropy ratio) were considered in the simulation for different values of the reservoir and caprock. It should be pointed out that the permeability anisotropy is considered because of the heterogeneity of the formation rock, which was usually ignored in many simulation studies. For sandstone, which is the reservoir rock in the CCS unit, the permeability anisotropy ratio (K_v/K_h) ranges from 0.19 to 0.77 (Clavaud et al. 2008). The critical parameter interjection rate was also considered in this study. Based on the work conducted by Rutqvist and Tsang (2002) and Bao et al. (2013), a linear downscaling injection rate was applied to harmonize the 2D geological model. The tested input parameters and their ranges used for sampling are shown in Tab. 3.2. The range of some parameters, such as the reservoir permeability, was slightly modified, while the other parameters in the geological model were taken from Rutqvist et al. (2010), as shown in Tab. 3.3.

Table 3.2 Tested input parameters and their ranges (Cao et al. 2020)

No.	Parameter	Min	Max	Ave	Source
1	Caprock Young's modulus [GPa]	5	45	25	modified from Rybacki et al. 2015
2	Caprock Poisson's ratio [GPa]	0.05	0.32	0.185	Gercek 2007
3	Caprock Biot coefficient [-]	0.6	1.0	0.8	Newell et al. 2017
4	Reservoir Young's modulus [GPa]	5	40	22.5	Hart and Wang 1995; Palmström and Singh 2001; Rutqvist and Tsang 2002
5	Reservoir Poisson's ratio [-]	0.05	0.4	0.225	Gercek 2007
6	Reservoir Biot coefficient [-]	0.5	1.0	0.75	Newell et al. 2017
7	Caprock permeability [Log10 m^2)	-21	-19	-20	Rutqvist et al. 2010
8	Reservoir horizontal permeability [Log10 m^2]	-14	-11.52	-12.76	The main CCS projects worldwide (Tab. 3.1)
9	Reservoir permeability anisotropy ratio [K_v/K_h]	0.19	0.77	0.48	Clavaud et al. 2008
10	Injection rate [kg/s]	0.05	0.3	0.175	The main CCS projects worldwide (Tab. 3.1)

Table 3.3 The additional parameters used for simulation (Cao et al. 2020; Rutqvist et al. 2010)

Layer	Overburden	Caprock	Reservoir	Base rock
Young's modulus [GPa]	1.5	/	/	20
Poisson's ratio [-]	0.2	/	/	0.15
Biot coefficient [-]	1	/	/	1
Porosity [-]	0.1	0.01	0.17	0.01
Residual gas saturation [-]	0.05	0.05	0.05	0.05
Residual liquid saturation [-]	0.3	0.3	0.3	0.3
Van Genuchten (1980), P_0 [KPa]	19.9	621	19.9	621
Van Genuchten (1980), [m]	0.457	0.457	0.457	0.457

The Quasi-Monte Carlo method is applied to sample the parameters over their stipulated ranges (see Tab. 3.2). Quasi-Monte Carlo is a Quasi-random number generator, which can generate highly uniform samples of the unit hypercube. Actually, the Quasi-Random sequences is not random number, which is generated by a certain function. There are three Quasi-random sequences, including Halton sequences, Latin hypercube sequences, and Sobol sequences (Bratley and Fox 2003). The Sobol sequences is generated by the Sobolset functions, which enables specify both a Leap property and a Skip of a quasi-random sequence. In addition, the scramble method of the sobolset classes enables a variety of scrambling techniques that can reduce correlations and improve uniformity simultaneously. On account of that, the Quasi-Monte Carlo method has a high performance in terms of randomly sampling high dimensional parameter spaces without gaps or clumping, which is often used to generate multi-dimensional random number (Bao et al. 2013; Hou et al. 2014).

By using the Quasi-Monte Carlo method, the sampling data was firstly produced in a range from 0 to 1 using the Sobol sequence generated by Sobolset functions, which could reduce the correlations and simultaneously improve the uniformity (Bratley and Fox 2003). After that, the sampling data were scaled to the actual input parameter ranges. Finally, 256 data points were generated in the first step. The typical data points such as those for which the permeability and injection rate approach opposite extremes were excluded from the CCS operation based on the reasonableness of the data. Finally, the remaining 238 data points are shown in Fig. 3.5 using paired scatter plots. It can be seen that the sampled data points are uniformly and randomly distributed in the parameter space without any replicated values of the input parameters.

Figure 3.5 Paired scatter plots of the input parameters (Cao et al. 2020)

Using the THM coupled simulator TOUGH2MP(TMVOC)-FLAC3D, CO_2 was injected through a well at a constant rate for 10 years in the representative generic model based on the parameters obtained from every data point in Fig. 3.5. During this period, the reservoir and caprock fluid pressure are recorded at the key points C and D respectively, for further distance correlation analysis as well as the training of the SVR surrogate model. Similarly, the vertical displacement at the key points A, B, and, C are recorded to characterize the ground surface uplift, and formation deformation at the top and bottom of the caprock (see in Fig. 3.2).

3.3 Correlation of formation response with parameter uncertainty

After the numerical simulation was conducted, the formation responses including the formation pressure and deformation were obtained for every parameter data point in Fig. 3.5. The distance correlation coefficient with high performance that was used to characterize the dependence of two vectors is then introduced as an indicator to determine the key parameters of different formations responses.

3.3.1 Correlation of pressure change with parameter uncertainty

The pressure changes of both the reservoir and the caprock can potentially re-activate the faults in the caprock, further compromising its integrity. Therefore, a pressure change is an essential indicator for the integrity assessment of caprock in CCS operation. Based on the standard at the initial pressure, only a positive pressure increment can be measured in this work under constant and continuous CO_2 injection. During the process of CO_2 injection, the distance correlation of different parameters along with the pressure increment in the reservoir and the caprock is shown in Fig. 3.6a and b, respectively.

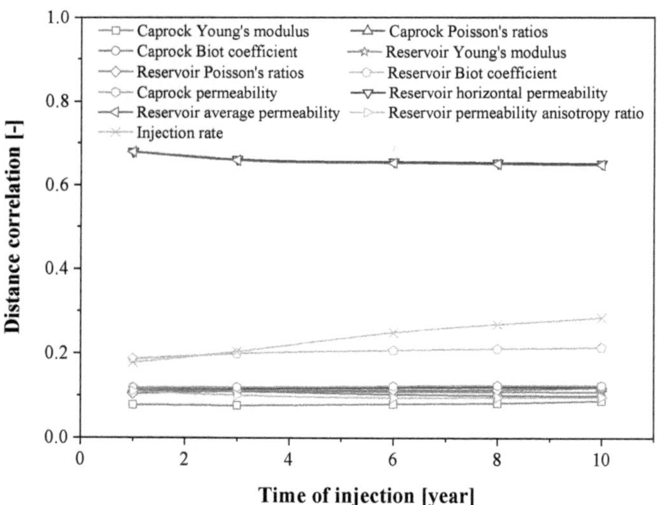

(a) Reservoir fluid pressure increment

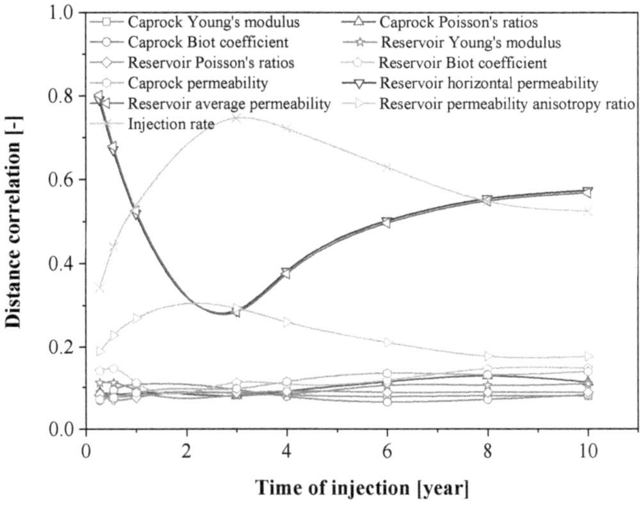

(b) Caprock pressure increment

Figure 3.6 Distance correlation for the formation pressure increments with the research parameters in the process of CO_2 injection (Cao et al. 2020)

For a pressure change in the reservoir, it can be seen from Fig. 3.6a that the distance correlation of the reservoir permeability (containing reservoir horizontal permeability and reservoir average permeability) reaches approximately 0.65, indicating a strong correlation between the reservoir permeability and the reservoir fluid pressure change. This demonstrates that the reservoir permeability acts as the critical parameter for the reservoir fluid pressure change, followed by the injection rate and the reservoir Biot coefficient. This indicates that the effect of the reservoir Biot coefficient should not be ignored, as has been the case in previous research (Bao et al. 2013), but should be taken into consideration. The next important parameter is the caprock permeability, which is consistent with the results for the generalized cross-validation and the analysis of the variance method. It should be pointed out that the ranks of Young's modulus and Poisson's ratio are not in accordance with the results result from that the impact on reservoir fluid pressure change is too weak to be detected.

As can be seen in Fig. 3.6b, the injection rate and reservoir permeability are two critical parameters for the formation fluid pressure change in caprock. It should be mentioned that their performance on caprock pressure increment shows a clear temporal difference as well as an inverse correlation. The permeability-dominated effect is clearly significant within the first year, given that the distance correlation reached 0.8 initially. The relatively low impact of the injection rate resulted in a low total volume of injected CO_2. With continuous injection, the distance correlation of the permeability decreases. Inversely, the gradually increasing distance correlation of the injection rate reaches a maximum value of 0.75 after 3 years of injections and dominates during the period from 1 to 8 years. During this time period, the accumulated volume of CO_2 has a significant effect on the formation fluid pressure change in the caprock. After approximately 8 years of injection, the accumulated volume of injected CO_2 reaches a high level. The pressure change in the caprock is then mostly dependent on the migration ability of the accumulated CO_2 in the reservoir. Therefore, the reservoir permeability once again has a dominant influence after 8 years of injection.

In addition to the injection rate and reservoir permeability, the reservoir permeability anisotropy ratio is a secondary parameter. It depicts a very similar but low effect compared to the injection rate. The distance correlation of the permeability anisotropy ratio shows an initial increase followed by a decrease, which is the opposite of the reservoir permeability. As shown in Tab. 3.2, the permeability anisotropy ratio is defined as the ratio between the vertical and horizontal permeability. Initially, the accumulated volume of injected CO_2 is not enough to affect the caprock. In this case, the preferential migration direction (the permeability anisotropy ratio) has a limited effect on the pressure change in the caprock. When the accumulated CO_2 invades the caprock, it is apparent that the pressure change is affected by a higher vertical permeability, namely, a higher permeability anisotropy ratio. This can promote CO_2 flow in the vertical direction. However, this influence decreases with a continuous injection result from that the pressure change is then determined mainly by the migration ability instead of the migration directions, with the accumulated CO_2 reaching a relatively high level. The other parameters such as the Biot coefficient and Poisson's ratio that are presented in Fig. 3.6b have a negligible effect on the pressure change in the caprock.

3.3.2 Correlation of formation deformation with parameter uncertainty

Formation deformation is another key point in characterizing the formation response during CO_2 injection. The deformation (vertical displacements) was recorded exactly over the injection point and analyzed at the ground surface as well as the top and bottom of the caprock. The correlation between the recorded displacements for different parameters is shown in Fig. 3.7.

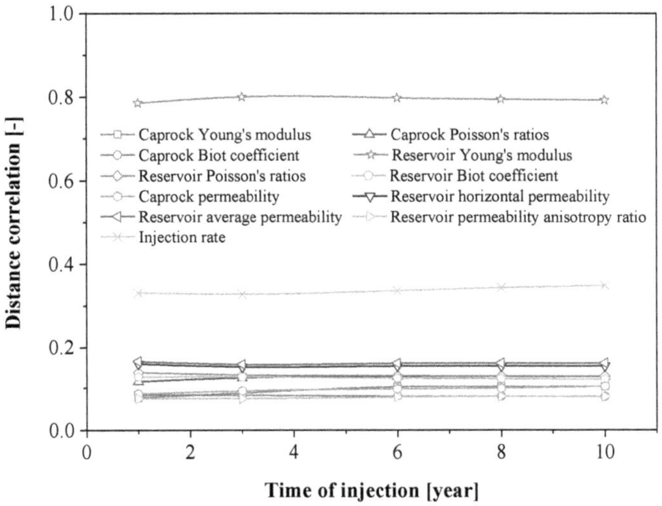

(a) Ground surface uplift

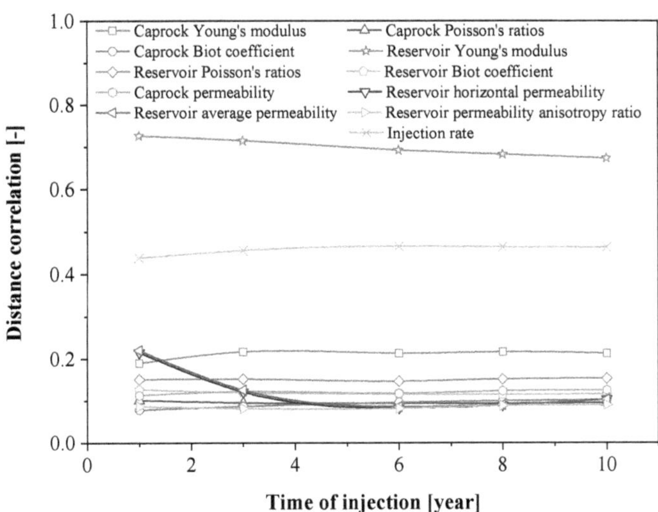

(b) Vertical displacement at the top of the caprock

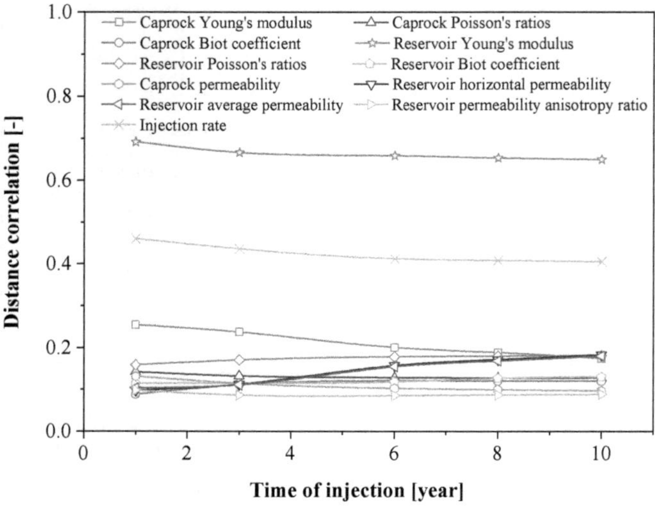

(c) Vertical displacement at the bottom of the caprock

Figure 3.7 Distance correlation for the formation vertical displacement based on the research parameters for the period of injection (Cao et al. 2020)

As depicted in Fig. 3.7, the reservoir Young's modulus and injection rate are the two most important parameters that affect the vertical displacement in all three positions at the ground surface, the top of the caprock and the bottom of the caprock. Wherein, the reservoir Young's modulus shows a much higher distance correlation for all three of the vertical displacements in the range from 0.65 to 0.80. Thus, this parameter is the most crucial factor in the determination of the deformation of the reservoir, caprock and overburden formations. Fig. 3.7 depicts that the injection rate is the next most important parameter for vertical displacement, which has a distance correlation in the range of 0.33 to 0.47 with vertical displacement. This indicates a moderate correlation between the injection rate and deformation. It should be mentioned that this order of importance for the different parameters with respect to formation deformation is different from the results obtained by Bao et al. (2013), in which the importance decreases with the parameters from reservoir permeability and injection rate, to the reservoir Young's modulus. The reason account for that is as follows. In their work, both the reservoir permeability and injection rate are considered over a much wider range of 10^{-15} to 10^{-12} m^2 and 0.01 to 1.5 million tons per year, respectively. However, a reservoir with a permeability of 10^{-15} m^2 is too tight to be used for CO_2 storage because of the low injectivity and storage capacity (Bachu et al. 2009; Raza et al. 2016). In addition to this, the injection scheme with a low injection rate of 0.01 million tons per year cannot be regarded as a commercial or demonstrative CCS operation (Michael et al. 2010). According to previous studies (Homma and Saltelli 1996; Li et al. 2012), the range of parameters represents our knowledge or lack thereof, and plays a crucial role in sensitivity analysis. For instance, the results acquired through sensitivity analysis may differ depending on the different ranges of the parameters. Therefore, it is inferred that the importance of the reservoir permeability and the injection rate may be magnified under a much wider range in Bao et al. (2013)'s work.

It should be emphasized that the caprock Young's modulus also has a small effect on vertical displacement in caprock formation. The other parameters with distance correlation lower than 0.2 show a substantially weaker relationship. This is result from that the other parameters have an extremely limited impact on the deformation of formation.

3.3.3 Correlation of the risk factor with the formation response

Brittleness is commonly regarded as an indicator in characterizing the possible failure features of rocks in oil and gas exploration (Jin et al. 2014; Rickman et al. 2008; Zhang et al. 2016). In this instance, it is used to evaluate the integrity of the caprock. As introduced by Rickman et al. (2008), the brittleness of the shale could be calculated based on the geomechanical properties of the rock, such as the Poisson's Ratio and Young's Modulus. The mathematical formula for brittleness is expressed as follows:

$$E_{\mathrm{Brit}} = (E - E_{\mathrm{min}}) / (E_{\mathrm{max}} - E_{\mathrm{min}}) \times 100 \tag{3.20}$$

$$\mu_{\mathrm{Brit}} = (\mu_{\mathrm{max}} - \mu) / (\mu_{\mathrm{max}} - \mu_{\mathrm{min}}) \times 100 \tag{3.21}$$

$$B_{\mathrm{Brit}} = 0.5 E_{\mathrm{Brit}} + 0.5 \mu_{\mathrm{Brit}} \tag{3.22}$$

where E_{Brit} denotes the normalized Young's Modulus, μ_{Brit} represents the normalized Poisson's Ratio, and B_{Brit} is the brittleness.

Based on the typical range of Young's Modulus and the Poisson's Ratio of caprock (Tab. 3.2), the brittleness is calculated and depicted in Fig. 3.8 in comparison with these two parameters. It can be clearly seen that the brittleness has a positive correlation with Young's modulus, and a negative correlation with Poisson's ratio. The brittleness of caprock was calculated based on the 238 sampled data points (Fig. 3.5), and their superimposition is depicted in Fig. 3.8. The well-dispersed sampling data points provide a good foundation to evaluate the potential risk of the caprock fully and efficiently.

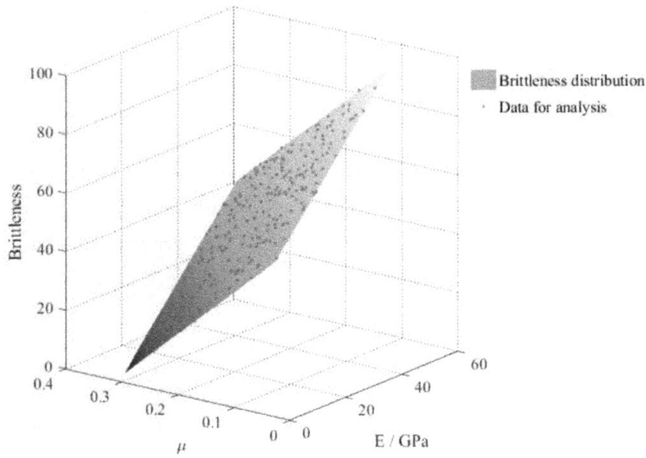

Figure 3.8 Brittle distribution of the caprock (Cao et al. 2020)

The mechanism of stress increment on the loss of caprock integrity was revealed by Dempsey et al. 2014). A higher stress increment is more likely to cause the reactivation of the current critically stressed fault, further compromising the caprock's integrity. Due to the constant weight of the overlaying caprock, the vertical stress is kept constant. As a result, the horizontal stress increment induced by the pressure change can be derived using pore-elastic theory, referring to the work of Hou et al. (2009). It can be illustrated as follows:

$$\Delta\sigma_{\mathrm{H/h}} = \alpha \frac{1-2v}{1-v} \Delta p \tag{3.23}$$

$$\Delta\sigma_z = 0 \tag{3.24}$$

where $\Delta\sigma_{\mathrm{H/h}}$ denotes the change in the total horizontal stress, α represents the Biot coefficient, μ is the Poisson's Ratio, Δp denotes the pore pressure change, and $\Delta\sigma_z$ represents the change of the total vertical stress.

Transform the stress change to be dimensionless as follows:

$$\Delta\sigma' = (\Delta\sigma - \Delta\sigma_{\mathrm{min}}) / (\Delta\sigma_{\mathrm{max}} - \Delta\sigma_{\mathrm{min}}) \tag{3.25}$$

To combine the impact of brittleness and stress increment on the caprock integrity, a risk factor defined as the product of brittleness and stress increment is introduced in this work. This is represented as:

$$R = E_{\mathrm{Brit}} \times \Delta\sigma' \tag{3.26}$$

According to the expression of the risk factor, the physical properties of the caprock, and the corresponding stress conditions were considered. As such, it can be used to characterize the instability of caprock in the CCS unit. Obviously, the higher the value of the risk factor, the higher the instability of the caprock. It is apparent that a greater risk factor will be obtained under higher levels of brittleness or a significant stress increment. The distance correlation of the risk factor with the measured formation response, such as formation pressure changes and displacements, is analyzed and depicted in Fig. 3.9.

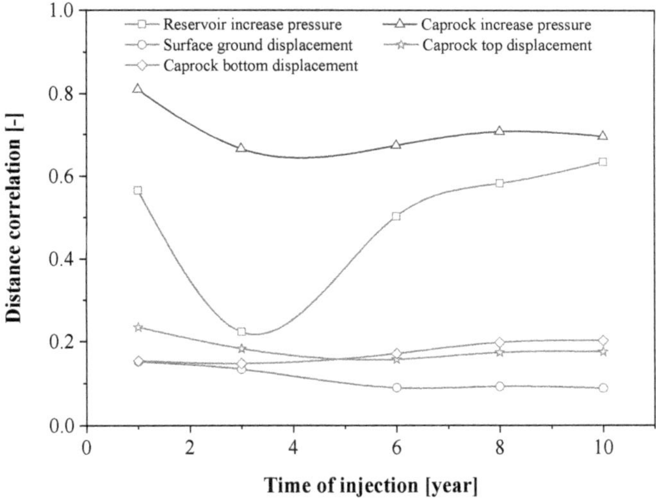

Figure 3.9 Distance correlation of the risk factor with the measured parameters during the period of injection

It can be seen that the pressure change in the caprock is the most critical factor in determining the risk factor, followed by the pressure change in the reservoir, for which the distance correlation has an average value of approximately 0.5. These two pressure changes have a close correlation with the risk factor. Fig. 3.9 shows that the vertical displacements at the top and bottom of the caprock and the ground surface only shows an average distance correlation of approximately 0.19, 0.18, and 0.11 with the risk factor, respectively. This demonstrates a relatively weak correlation between all three displacements with the risk factor. Therefore, it can be concluded that the pressure change is a valuable indicator for the integrity assessment in caprock compared with the formation displacements, which is further supported by the results conducted by Chen et al. (2018).

3.4 Surrogate model based on SVR

With the help of the measured formation pressure change and displacements and the corresponding parameters, the specific surrogate model can predict the formation response under specific parameters involving trained by using SVR.

3.4.1 SVR surrogate model for formation pressure

The sampling dataset obtained from the simulation was separated into two parts. Specifically, the dataset was divided into 200 and 38 datasets, respectively. The dataset with the sampling data of 200 was used to obtain a trained SVR surrogate model, which could then be verified by using the sampling data of 38. The coefficient of determination (R^2) was introduced as a measurement of goodness of fit to compare the predicted value and the actual value obtained from the computational simulation (Pan et al. 2016; Wriedt et al. 2014). If the value of R^2 was close to 1, indicating that the ratio of the sum of the squares for the residuals relative to the total sum of squares near zero. This also implies that the SVR surrogate model can predict the formation response due to CO_2 injection with very high accuracy.

Additionally, the normalized root mean square error (NRMSE) was used to examine the efficacy of the SVR surrogate model for predictions. The NRMSE can be expressed as (Wriedt et al. 2014):

$$\mathrm{NRMSE} = \frac{\mathrm{RMSE}}{Y_{\mathrm{pred,max}} - Y_{\mathrm{pred,min}}} \times 100\% \tag{3.27}$$

where $Y_{\mathrm{pred,max}}$ denotes the maximum value of the predicted outcome, and $Y_{\mathrm{pred,min}}$ represents the minimum value of the predicted outcome. It is apparent that the smaller the value of NRMSE, the higher the accuracy of the predicted results obtained using the trained SVR surrogate model. The root mean square error (RMSE) is calculated in the following equation:

$$\mathrm{RMSE} = \sqrt{\frac{\sum_{i=1}^{n} \left(Y_{\mathrm{pred,i}} - Y_{\mathrm{sim,i}}\right)^2}{n}} \tag{3.28}$$

where Y_{pred} represents the predicted value, and Y_{sim} denotes the simulated value.

The predicted value of the training dataset is shown in Fig. 3.10a and 3.11a together with the numerically obtained actual formation fluid pressure increment in the reservoir and the caprock, respectively. It can be clearly seen that the predicted value of the fluid pressure increments for both the reservoir and the caprock shows a perfectly linear correlation with the numerically determined actual value. This demonstrated that the value of R^2 were up to 0.9952 and 0.9842, and the NRMSE values were only 1.66 % and 2.88 %, respectively. This means that these two SVR surrogate models for the reservoir fluid pressure and caprock fluid pressure were well-trained. Using the remaining 38 test datasets, the trained SVR surrogate models could be verified. As can be seen in Fig. 3.10b and 3.11b, the data points with a highly accurate predicted data are much closer to the diagonal than other data points. Especially, if the predicted data is equal to the measured one, the corresponding data point will be located exactly on the diagonal of the figure. As depicted in Fig. 3.10b and 3.11b, the predicted data points are located in a narrow band range surrounding the diagonal, demonstrating the availability of the trained SVR surrogate model. The corresponding R^2 and NRMSE were calculated to be approximately 0.9 and 0.1, respectively. Therefore, the trained SVR surrogate model can predict the fluid pressure change with acceptable accuracy.

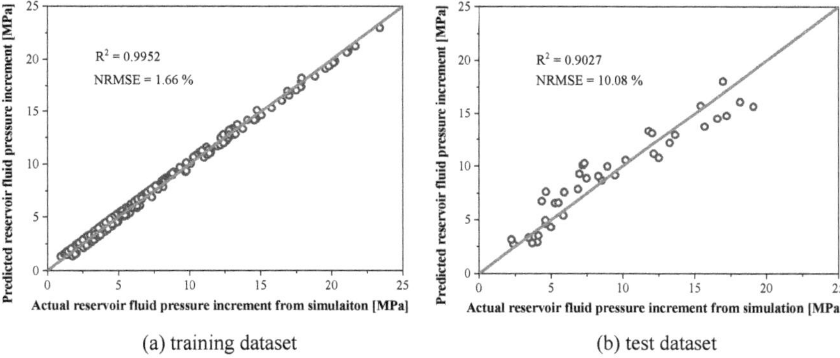

(a) training dataset (b) test dataset

Figure 3.10 Predicted and actual reservoir fluid pressure increment at 10 years (Cao et al. 2020)

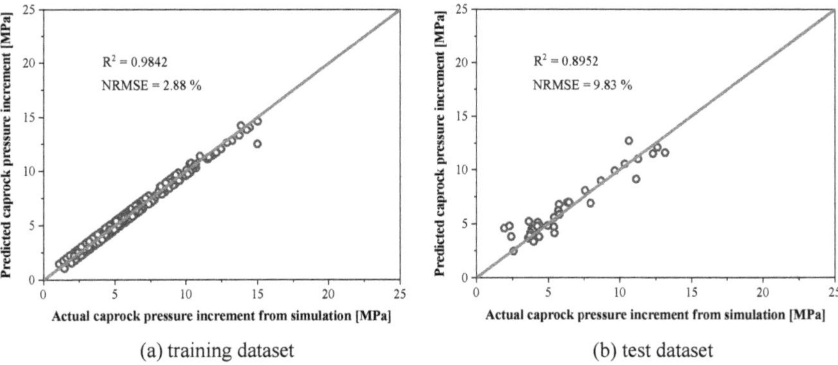

(a) training dataset (b) test dataset

Figure 3.11 Predicted and actual caprock pressure increment at 10 years (Cao et al. 2020)

It should be pointed out that the prediction accuracy for the test dataset is less than that of the training dataset. This is result from that the information related to the training dataset is included in the SVR surrogate model. However, the characteristics of the test dataset are not included. In addition, the sample size is not very large. If a larger dataset is applied for training and testing, a superior prediction accuracy can be achieved.

3.4.2 SVR surrogate model for formation deformation

Similar to the training of the SVR surrogate model for formation pressure change, the SVR surrogate model of the corresponding formation displacement was trained on the division of 200 training datasets. Subsequently, the predicted formation displacement of the training dataset is compared to the numerically determined actual displacements at the ground surface, the top of the caprock, and the bottom of the caprock as presented in Fig. 3.12a, 3.13a, and 3.14a, respectively. Clearly, the SVR surrogate models of all the three displacements were efficiently trained. It is apparent that almost all the data points are located on the diagonal of the figure. This is evident given that all three R^2 values exceed 0.99, showing a high prediction accuracy. The corresponding NRMSE values are 9.35%, 5.21%, and 0.45%, respectively.

Likewise, the trained SVR surrogate model was tested by using the remaining 38 datasets. The results are depicted in Fig. 3.12b, 3.13b, and 3.14b, respectively. In comparison with the pressure change, the predicted displacements obtained from the trained SVR surrogate model show relatively high accuracy. This can be determined because the R^2 values of the test datasets are larger than 0.92, and the corresponding NRMSE values are smaller than 8 %. It is worth to mention that the R^2 value increases whereas the NRMSE decreases along the predicted value in the test dataset from the ground surface, to the top and bottom of the caprock. This means that the SVR surrogate model of the formation displacements at the bottom of the caprock performs better than the other two models.

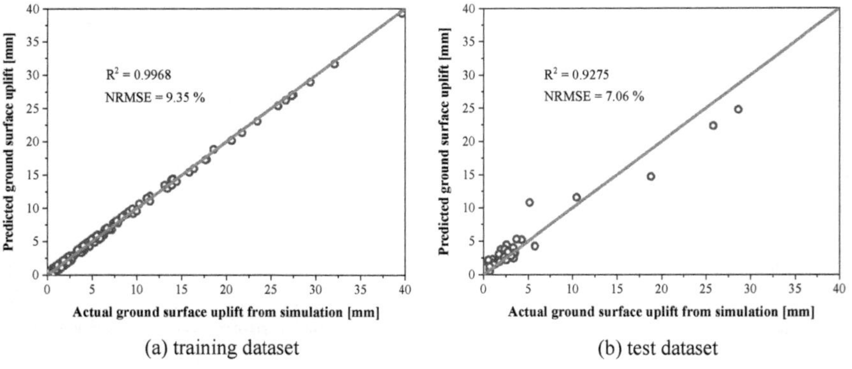

Figure 3.12 Predicted and actual ground surface uplift (Cao et al. 2020)

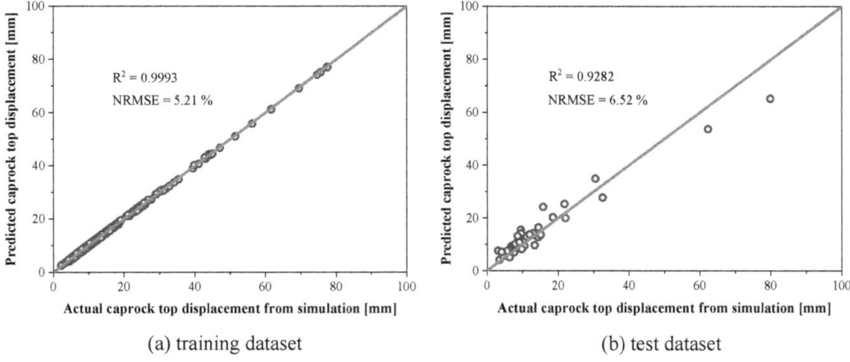

Figure 3.13 Predicted and actual displacement at the top of the caprock (Cao et al. 2020)

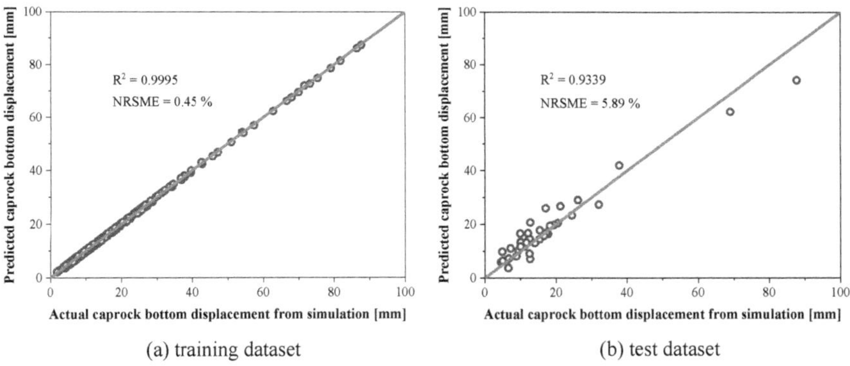

Figure 3.14 Predicted and actual displacement at the bottom of the caprock (Cao et al. 2020)

3.5 Application of the SVR surrogate model to sensitivity analysis

Based on the well-trained SVR surrogate model, the corresponding formation response after 10 years of CO_2 injection is predicted again in a much denser 2D sampled space of key parameters, and the effect of the other secondary parameters is excluded using their average values in Tab. 3.2. A sensitivity analysis of the key parameters corresponding to the predicted formation response is then conducted to examine how the key parameters impact the formation responses.

3.5.1 Sensitivity analysis of the formation pressure

To systematically investigate the role of the parameters, the key parameters of pressure change including the reservoir permeability and the injection rate determined based on previous distance correlation analysis were resampled into 512 data points over a given range by using the Quasi-Monte Carlo method. The average values of the other parameters were used (see Tab. 3.2). Based on these data points, the fluid pressure changes after 10 years of CO_2 injection were predicted using the well-trained SVR surrogate model in the reservoir and the caprock, respectively. The results are summarized in Fig. 3.15 against the key parameters, i.e., reservoir permeability and injection rate.

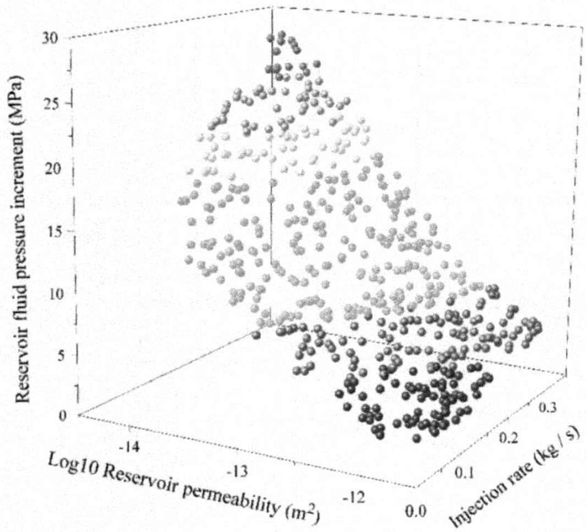

(a) Reservoir fluid pressure increment

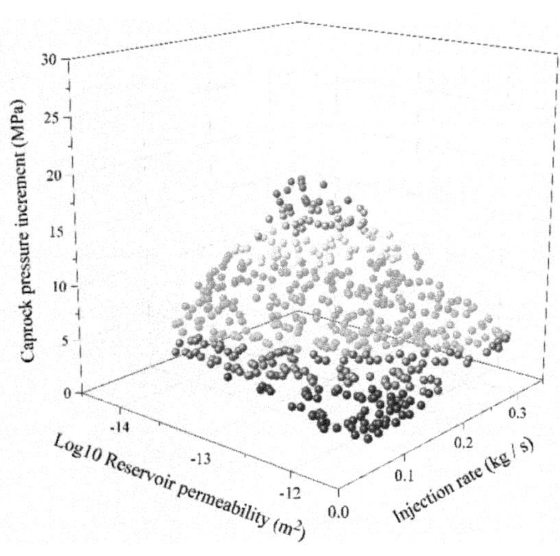

(b) Caprock pressure increment

Figure 3.15 The formation pressure variation under different reservoir permeability and injection rates (Cao et al. 2020)

As can be seen in Fig. 3.15a, the reservoir fluid pressure increment has a positive correlation with the injection rate but a negative correlation with the reservoir permeability. The predicted data point determined surface is inclined along with the reservoir permeability axis. This means that the reservoir fluid pressure increment increases dramatically with the decrease in the reservoir permeability, whereas it increases relatively slowly with the increase in the injection rate. Therefore, it can be concluded that the reservoir fluid pressure change is more sensitive to the reservoir permeability compared with the injection rate, which is consistent with the results obtained from the distance correlation analysis.

Fig. 3.15b shows the predicted data of the pressure change in the caprock against the reservoir permeability and injection rate. It exhibits a very similar but lower tendency compared to that of the pressure change in the reservoir. In addition, the degree of inclination of the surface along the reservoir permeability and the injection rate are approximately equal. This can result in an identical influence of the reservoir permeability and the injection rate on the caprock pressure change. Moreover, it is also in accordance with the results obtained from the previous distance correlation analysis, in which the distance correlation at 10 years was 0.57 and 0.52 for the reservoir permeability and injection rate, respectively (Fig. 3.6b).

Comparing Fig. 3.15a and 3.15b, a similar tendency of the fluid pressure increments with respect to the reservoir permeability as well as the injection rate is observed in the reservoir and the caprock, wherein the fluid pressure increment in the reservoir is higher. Specifically, the maximum values reached 27.5 MPa in the reservoir and 15 MPa in the caprock under low permeability and high injection rate conditions. Therefore, it is suggested that a low injection rate should be used in a reservoir with low permeability to avoid extreme pressure in both the reservoir as well as the caprock.

In the forementioned sensitivity analysis, a total of 512 data points were generated for the purpose of covering the whole range of the two key parameters (reservoir permeability and injection rate) and obtain the response surface of reservoir fluid pressure increment. The physical model needs to be simulated for 512 times to acquire the results, which cost several days. However, the results can be obtained by the surrogate model within only several seconds, showing the efficiency of the SVR surrogate model.

3.5.2 Sensitivity analysis of the formation deformation

The formation deformation (ground surface uplift, the top and bottom of the caprock) is predicted in the same way according to the SVR surrogate model based on its own key parameters including the injection rate and the reservoir Young's modulus. A total of 512 new randomly sampled data points were applied to predict the three vertical displacements after 10 years of CO_2 injection by using the trained SVR surrogate model. The results are depicted in Fig. 3.16 for the three vertical displacements.

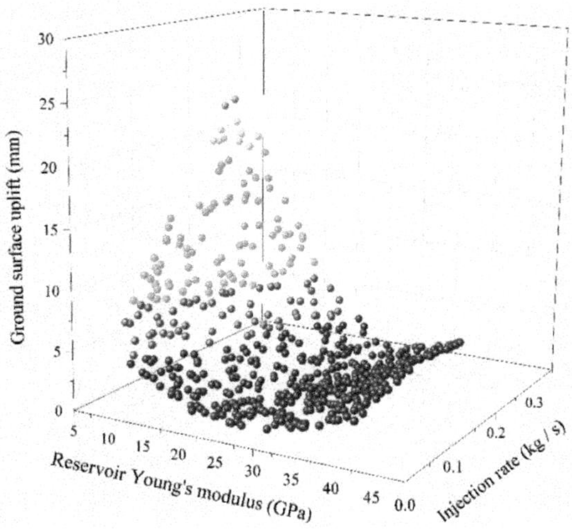

(a) Ground surface uplift

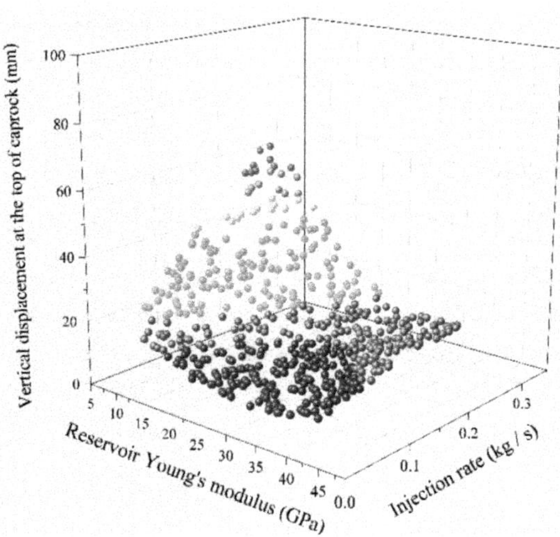

(b) Vertical displacement at the top of the caprock

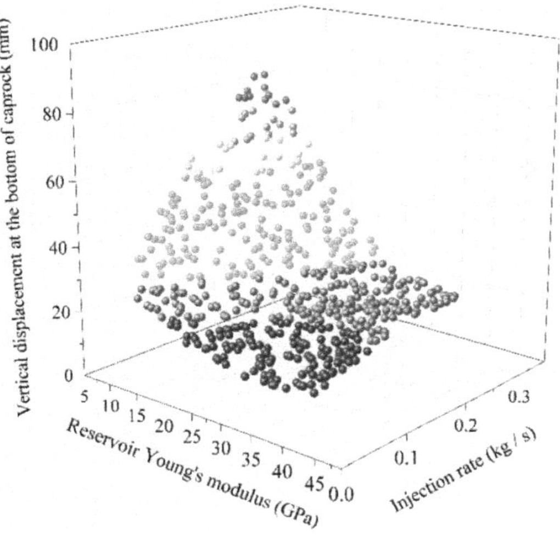

(c) Vertical displacement at the bottom of the caprock

Figure 3.16 The formation deformation variation under different reservoir Young's modulus and injection rates (Cao et al. 2020)

Fig. 3.16 exhibits that all the predicted data point determined surfaces are inclined along the reservoir Young's modulus axis, indicating that the vertical displacement in all three positions is more sensitive to the reservoir Young's modulus compared with that of the injection rate. This is in accordance with the results obtained for the distance correlation analysis given that the reservoir Young's modulus is more closely correlated to the formation vertical displacement at all three positions (Fig. 3.7). Fig. 3.16 also demonstrates that the formation vertical displacement reacts slowly when the reservoir Young's modulus is over 25 GPa. In contrast, the formation vertical displacements for all three positions are more sensitive to the reservoir Young's modulus for values under 25 GPa. It is probable that the reservoir Young's modulus of a CCS project will be less than 25 GPa, such as Young's modulus for the In Salah project, which was only 6 GPa (Rutqvist et al. 2010). This demonstrates the importance of the reservoir Young's modulus on formation deformation.

Overall, the vertical displacement at the ground surface, top, and bottom of the caprock also shows a comparable tendency relative to the reservoir Young's modulus and the injection rate, which increases with the decrease in Young's modulus, and the increase in the injection rate. The maximum formation vertical displacement of approximately 80 mm is found at the bottom of the caprock associated with a low Young's modulus and high injection rate. Consequently, the formation displacement, especially at the bottom of the caprock, is supposed to be carefully assessed associated with a low Young's modulus and a high injection rate.

In the forementioned sensitivity analysis, a total of 512 data points were produced to cover the whole range of the two key parameters (reservoir Young's modulus and the injection rate) and then acquire the vertical displacement of the formation. Thus, a total of 512 computational simulations need to be implemented if the physical model was used, which cost a very long time. However, the results can be acquired by the SVR surrogate model in only several seconds, demonstrating the efficiency of the SVR surrogate model. In general, the SVR surrogate models can be used to save computational costs particularly for abundant simulation work.

3.6 Summary

Using the TOUGH2MP(TMVOC)-FLAC3D simulator, a numerical simulation for hundreds of sampled data was conducted for results generated with the help of the Quasi-Monte Carlo method. Based on the simulation results, the general role of different geomechanical and hydrogeological parameters was analyzed in response to CO_2 injection by using distance correlation. This can be used to guide time and effort spent in mitigating the uncertainty in these parameters to obtain trustworthy model forecasts and risk assessments in CCS projects. In this work, within the parameter ranges in Tab. 3.2 for a given generic configuration geological model, the reservoir permeability and the injection rate are the two key factors in determining the pressure change. Moreover, the reservoir Young's modulus plays the most important role in formation deformation including vertical displacement. The pressure change shows a much closer correlation with the risk factor in comparison to the formation deformation, demonstrating the importance of pressure change in the risk assessment of the caprock.

Based on the machine learning approach in SVR, the SVR surrogate model was well-trained on the base of the data regarding simulated results and its reliability was verified by using the test dataset. Thereafter, the formation response including the formation pressure change as well as formation deformation, were predicted by using the trained SVR surrogate model on the basis of the re-sampling data, over the aforementioned key parameters. The rules gained from the predicted results and the distance correlation analysis were confirmed on the base of a comparison to each other. The methods and working scheme used in this work could be applied for the uncertainty analysis of other geological operations.

4 CO_2 storage with impurities associated with enhanced gas recovery in depleted gas reservoirs

In this chapter, to investigate the suitability of co-injecting impurities with CO_2 into depleted gas reservoirs, three different mole fractions of N_2 and O_2 in the mixture gas were used to represent the flue gas and been injected into a typical depleted gas reservoir. The scenario of both dedicated for CO_2 storage and CO_2 storage with enhanced gas recovery (CSEGR) were taken into consideration. In the scenario of dedicated for CO_2 storage, the gas was injected into the depleted gas reservoir until the average reservoir pressure reach to the original pressure of the reservoir. In the scenario of CSEGR, CO_2-EGR was conducted firstly and then transformed for CO_2 storage. The effect of residual CH_4 content, reservoir temperature, residual water saturation, and injection rate on the performance of CO_2 storage and CSEGR were analyzed in detail. The main contents of this chapter have been prepared as a manuscript in the following research paper (Cao et al. 2020): Numerical modeling for CO_2 storage with impurities associated with enhanced gas recovery in depleted gas reservoirs.

4.1 Impurity gas related to CO_2 storage

4.1.1 Impurity level

The treated flue gas captured from large atmospheric CO_2 emission sources is generally a gas mixture of CO_2 with N_2, O_2, and other impurities such as Ar, SO_2, NO_x. According to the gas components in the oxyfuel flu gas, N_2 and O_2 dominate the mole fraction and have been considered as the impurities co-injected with CO_2 into the depleted gas reservoirs (Li et al. 2011). As shown in Tab. 4.1, the gas mixture with three different concentration of impurities were applied for co-injection, including the gas with low, medium, and high concentration of impurities (abbreviated as L Impu, M Impu, and H Impu respectively in the Figures), which corresponding the component of 98% CO_2 + 1% N_2 + 1% O_2, 91.5% CO_2 + 5.5% N_2 + 3% O_2, and 85% CO_2 + 10% N_2 + 5% O_2, respectively.

Table 4.1 Typical major components of oxyfuel flue gas

Components	CO_2 [%]	N_2 [%]	O_2 [%]	Total [%]	Reference
Low concentration of impurities	98	1	1	100	Modified from Li et al. 2011; Wang et al. 2012
Medium concentration of impurities	91.5	5.5	3	100	
High concentration of impurities	85	10	5	100	

4.1.2 Physical properties of the impurity-laden CO_2

Fig. 4.1 shows the phase envelopes of impurity-laden CO_2 that are calculated by using the SRK equation of state (Soave 1972). It can be seen that there are two-phase regions (the envelope) occurred due to the exits of impurities. Further, the phase envelope expands with the increasing impurity concentration. The injected CO_2 usually would be in the supercritical state at the reservoir temperature and pressure conditions. Fig. 4.1 demonstrates that the critical temperature and pressure of the impurity-laden CO_2 are altered compared with the pure CO_2. Specifically, the impurities would decrease the critical temperature while increases the critical pressure of the mixture gases. As can be seen in Tab. 4.2, the critical temperature of the injected gas with low, middle, and high concentration of impurities are 29.78, 25.27, and 20.18 °C, respectively. The corresponding critical pressure are 7.61, 8.45, and 9.38 MPa, respectively. This is result from the notable difference of the critical properties between CO_2 and the impurities as shown in Tab. 4.2. The critical temperature of N_2 and O_2 is far lower than that of CO_2, thus the N_2 and O_2 would decrease the critical temperature of the mixture gases, which has been verified by the results conducted by Wang et al. (2011). Furthermore, the lower critical temperature of mixture gases has an effect on its properties such as density in the supercritical states.

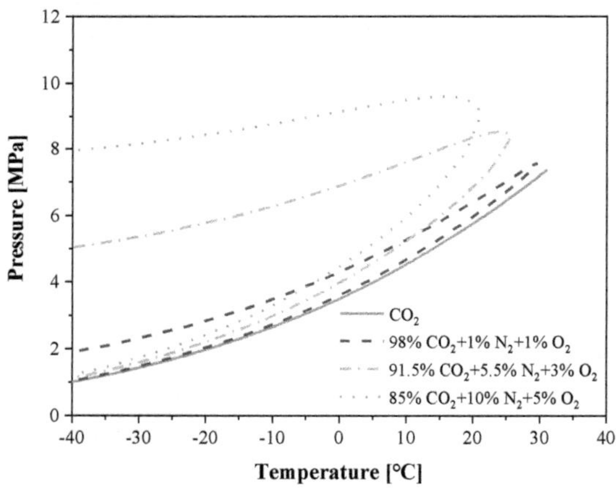

Figure 4.1 Calculated phase envelopes of the impurity-laden CO_2

Table 4.2 Critical properties of the gases

Gas	Critical temperature [°C]	Critical pressure [MPa]	Source
CO_2	31.04	7.38	Yaws 1999
N_2	-147.05	3.39	
O_2	-118.57	5.04	
98% CO_2 + 1% N_2 + 1% O_2	29.78	7.61	Calculated using SRK equation
91.5% CO_2 + 5.5% N_2 + 3% O_2	25.27	8.45	
85% CO_2 + 10% N_2 + 5% O_2	20.18	9.38	

Fig. 4.2a shows the density of CO_2 and the mixture gases with three concentration of impurities at the condition of 90 °C, which is a typical temperature for CO_2 storage. It can be clearly seen that the N_2 and O_2 reduce the density of the CO_2 stream significantly due to their lower compressibility, and such decreasing behavior augments with the growing concentration of impurities. The decrement of density increases firstly and then decreases with the increasing pressure. Specifically, the density difference

between the mixture gas with the pure CO_2 reaches the maximum value at approximately 22, 24, and 26 MPa for the gas with low, middle, and high concentration of impurities, respectively. On account of that the density of injected gases could be regarded an indicator for the storage capacity (Nicot et al. 2013), it can be inferred that the impact of pressure on the storage capacity increases firstly and then decreases along with the increasing pressure. Generally, the impact of impurities on the CO_2 storage capacity is dependent on the reservoir pressure condition and the concentration of the impurities. The effect of N_2 and O_2 on the viscosity of the mixture gases is similar to that of density as shown in Fig. 4.2b.

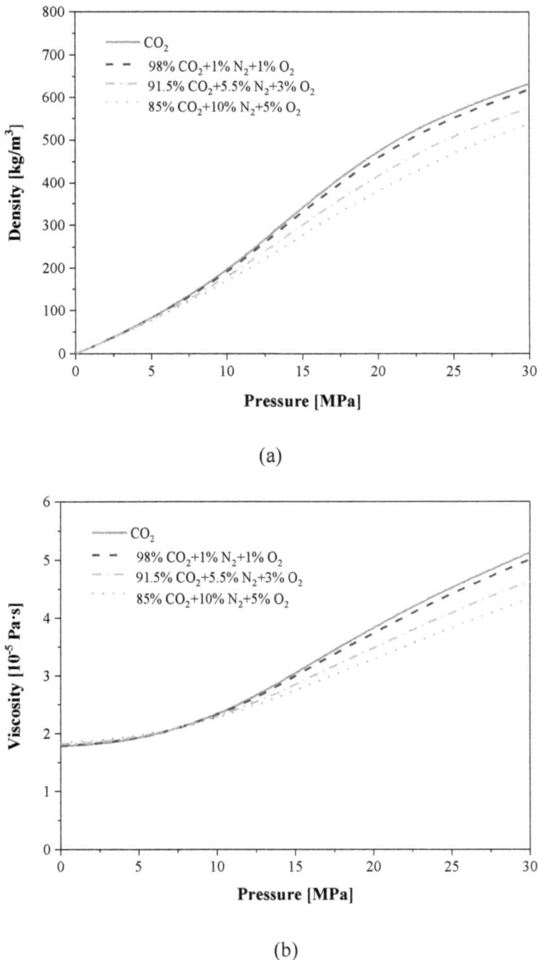

Figure 4.2 The (a) density and (b) viscosity of CO_2 and CO_2 mixtures as a function of pressure at the temperature of 90 °C

4.2 Generic model based on literature study

A typical depleted gas reservoir with a thickness of 50 m and located at the depth of 3000 m was used for simulation, which is modified from Zhang et al. (2017). The horizontal and vertical permeability of the reservoir are 10 and 5 mD, respectively. The reservoir porosity is 15% and the irreducible water saturation is 25%. A primary gas recovery of 80% was assumed, thus the reservoir pressure has been decreased from the original pressure of 30 MPa to the depleted pressure of 6 MPa. As depicted in Fig. 4.3a, a typical five-spot well pattern was applied in this study, wherein only one well is located at the lower formation for gas injection, and the other four wells are located at upper formation for gas injection. This well configuration is beneficial for the mitigation of the CO_2 breakthrough at the production well (Khan et al. 2012). In the symmetrical well pattern, only one-quarter model of the reservoir with confined boundary was selected for the simulation (Fig. 4.3b). The dimension of the reservoir model is 1000 m × 1000 m × 50 m. It was discretized into 25,000 rectangular elements. The Van Genuchten (1980) and Corey (1954)'s model for relatively permeability and the Van Genuchten (1980)'s model for capillary pressure were used for simulation. The detailed parameters applied for the model and simulation were summarized in Tab. 4.3.

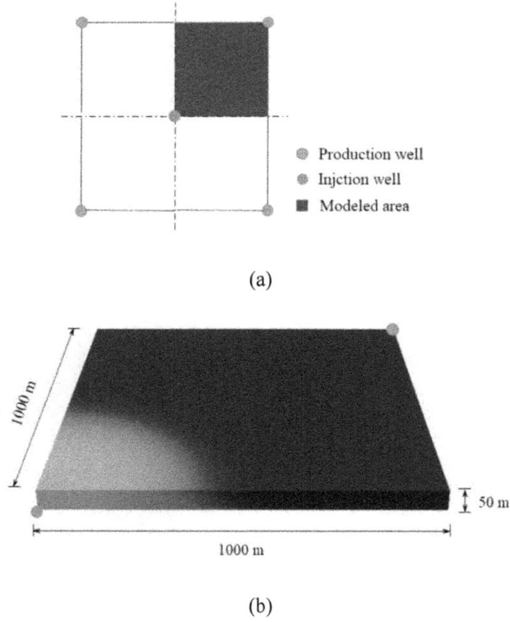

Figure 4.3 (a) Five-spot pattern depicting the CO_2 injection well and the production wells; (b) A quarter model applied for simulation (The production well only used in the process of EGR) (Adapted from Zhang et al. 2017)

Table 4.3 Main parameters of the reservoir simulation model (modified from Pruess and García 2002; Zhang et al. 2017)

Parameters	value
Length [m]	1000
Width [m]	1000
Thickness [m]	50
Porosity [-]	0.15
Horizontal permeability [mD]	10
Vertical permeability [mD]	5
Temperature [°C]	90
Original pressure [MPa]	30
Depleted pressure [MPa)	6
Residual water saturation [-]	0.25
Relative permeability model	
Liquid (Van Genuchten 1980)	
$k_{rl} = \sqrt{S^*}\left\{1-\left(1-\left[S^*\right]^{1/\lambda}\right)^{\lambda}\right\}^2$	$S^* = (S_l - S_{lr})/(1-S_{lr})$
S_{lr}: irreducible water saturation	S_{lr}=0.25
λ: exponent	λ=0.457
Gas (Corey 1954)	
$k_{rg} = (1-\hat{S})^2\left(1-\hat{S}^2\right)$	$\hat{S} = (S_l - S_{lr})/\left(1-S_{lr}-S_{gr}\right)$
S_{gr}: residual gas saturation	S_{gr}=0.05
Capillary pressure model (Van Genuchten 1980)	
$P_{cap} = -P_0\left(\left[S^*\right]^{-1/\lambda}-1\right)^{1-\lambda}$	$S^* = (S_l - S_{lr})/(1-S_{lr})$
S_{lr}: irreducible water saturation	S_{lr}=0
λ: exponent	λ=0.457
P_0: strength coefficient	P_0=19.61 KPa

4.3 Simulation schemes

4.3.1 Operation of dedicated for CO_2 storage

When the depleted gas reservoir is dedicated for CO_2 storage, the gas was injected into the reservoir without any gas production. The gas injection will continue until the average reservoir pressure recover to the original reservoir pressure (30 MPa).

4.3.2 Operation of CSEGR

There are two stages designed for the operation of CSEGR. The first one is gas injection associated with CH_4 production, and the second one is gas injection without production. During the process of CO_2-EGR, the gas is injected at the lower layer of the reservoir, whereas the CH_4 is produced at the upper layer of the reservoir away from the injection well. The gas production is terminated when the mole fraction of CH_4 in the produced gas is lower than 90%, which is taken from Al-Hasami et al. (2005). It should be mentioned that the CH_4 was produced under the condition of keeping the pressure on the production well boundary constant and equal to the initial pressure of the depleted reservoir (6 MPa). This production scheme was modified from Class et al. (2009)'s work. Regarding the second stage, the gas was injected through the well located at the lower layer without gas production. The operation of the project will continue until the average reservoir pressure recover to the original reservoir pressure of 30 MPa.

4.4 Dedicated for CO_2 storage

4.4.1 Effect of the gas recovery

According to the statistical data obtained by Laherrère (1997), the gas recovery is not closely related to the reservoir depth and mainly ranges from 70% to 90% as shown in Fig. 4.4. Based on this, the depleted gas reservoir with a gas recovery of 70%, 80% and 90% was applied for CO_2 injection, respectively. The main simulation results for the case of dedicated CO_2 storage with different primary gas recovery are illustrated in Fig. 4.5.

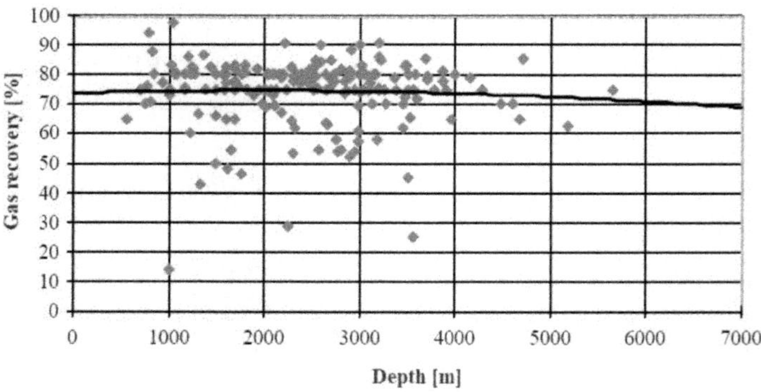

Figure 4.4 Gas recovery of world outside N. America (Laherrère 1997)

Obviously, more time is required for the reservoir with a higher gas recovery to recover to its initial pressure. For the reservoir with a certain gas recovery, the project duration decreases with the increasing of impurity gas concentration because of its lower compressibility compared with CO_2. For instance, the project durations for the reservoir with a gas recovery of 70% for the injection of pure CO_2 and the gas with low, middle, and high concentration of impurities are 8.13, 7.96, 7.37, and 6.85 years, respectively. The project durations are 10.74, 10.53, 9.74, and 9.05 years, respectively for the reservoir with a gas recovery of 90%. Moreover, it can be calculated that the difference of the project durations for the reservoirs with different gas recovery decrease with the growing impurity gas concentration, i.e., they are 2.61, 2.57, 2.37, and 2.2 years in previous cases, demonstrating the significant impact of the N_2 and O_2 on the pressure buildup of the depleted gas reservoir.

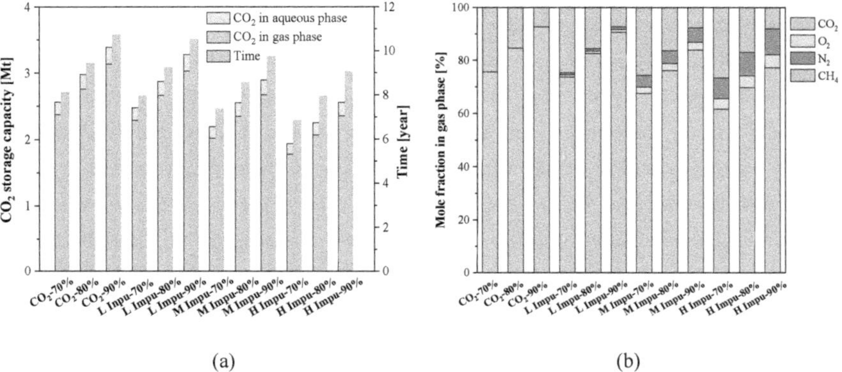

Figure 4.5 Main simulation results for the case of dedicated CO_2 storage with different primary gas recovery. (a) CO_2 storage capacity and the project duration; (b) Mole fraction of the gases in the reservoir

Fig. 4.5a shows that the CO_2 capacity increases from 2.56 to 3.39 Mt for the pure CO_2 when the gas recovery varies from 70% to 90%, i.e., an increment of 0.83 Mt CO_2 was obtained. These values are 0.80, 0.70, and 0.62 Mt for the injected gas with low, middle, and high concentration of impurities, respectively. This indicates that the effect of the residual CH_4 content on the CO_2 storage capacity decreases with the increasing impurity concentration. Regarding the reservoir with a primary gas recovery of 80%, the CO_2 storage capacity is 2.98 Mt for pure CO_2. These values are 2.88, 2.55, 2.25 Mt for the injected gas with low, middle, and high concentration of impurities respectively, which are 3.36%, 14.43%, and 24.50% lower than that of pure CO_2. It should be mentioned that these reduction degrees for the CO_2 storage capacity are almost identical to the corresponding value for the reservoirs with a gas recovery of 70% and 90%, indicating that the impact of the impurity gas on the CO_2 capacity is dependent on the impurity concentration of the injected gas. Considering that the impurity concentrations are 2%, 8.5% and 15% for the three different injected gases, it can be concluded that the impact of the N_2 and O_2 on the CO_2 capacity is proportional to the impurity gas concentration. The reason is that the density decrement of the mixture gas is approximately proportional to the concentration of impurities at the reservoir pressure of 30 MPa (Fig. 4.2a), which is the reservoir pressure corresponding the termination of the CCS operation. This is in accordance with the results obtained from Barrufet et al. (2010)'s study, which reveals that the storage capacity decreases proportionally to the concentration and the compressibility of N_2 when it is co-injected with CO_2.

Fig. 4.5a also shows that the solubility trapping capacity of CO_2 augments with the increasing of gas recovery. It increases from 0.19 to 0.25 Mt for pure CO_2 and 0.16 to 0.21 Mt for the gas with high concentration of impurities when the gas recovery varies from 70% to 90%. This may result from that the injected CO_2 is more likely to contact with the formation water and dissolve in the formation water associated with the reservoir with a high primary gas recovery and only less CH_4 left. For the reservoir with a certain gas recovery, obviously, the solubility trapping capacity of CO_2 decreases with the increases of impurity concentration. For example, for the reservoir with a gas recovery of 80%, the solubility trapping capacity of CO_2 are 0.22 and 0.18 Mt respectively for the gas with low and high concentration of impurity gas, which is consistent with Li and Jiang (2014)'s and Mahmoodpour et al. (2018)'s results associated with the impact of N_2 on the solubility trapping of CO_2. Generally, the role of solubility trapping on the CCS is limited due to the limited residual water saturation in the depleted gas reservoir, which captures only 7.3% to 8.2% of the total trapped CO_2 in above cases.

It can be seen from Fig. 4.5b that the mole fraction of CO_2 in gas phase increases with the increasing of gas recovery and the decreasing of impurity concentration. Regarding the reservoir with a gas recovery of 80%, the mole fraction of CO_2 in gas phase are 84.6%, 82.5%,76.1%, 69.7% for the pure CO_2, and the gas with low, middle, and high concentration of impurities. It can be calculated that the reductions are 2.1%, 8.5%, and 14.9%, which are approximately equal to the concentration of impurities in the injected gases, respectively. This situation also occurs for the reservoir with a gas recovery of 70% and 90%, which is result from the low solubility of N_2 and O_2 compared with CO_2 in the formation water.

4.4.2 Effect of the reservoir temperature

Assuming a ground temperature of 15 °C and a geothermal gradient from 20 to 33 °C/km for the subsurface that is taken from Wang et al. (2012), the effect of different reservoir temperature, i.e., 75, 90, and 114 °C respectively, on the performance of CO_2 storage with impurities in depleted gas reservoirs was investigated. The simulation results are shown in Fig. 4.6. As expected, the project duration decreases with the rising of impurity concentration and the temperature account for the low compressibility of N_2 and O_2 and the high pressure caused by high temperature, which is also demonstrated by Cao et al. (2020). Likely, the CO_2 storage capacity also decreases with the increasing of temperature. Specifically, it decreases from 3.28 to 2.57 Mt, 3.16 to 2.49 Mt, 2.79 to 2.21 Mt, and 2.46 to 1.97 Mt for pure CO_2, and the injected gas with low, middle, and high concentration of impurity gas respectively when the temperature increases from 75 to 114 °C. It can be calculated that the decrement of CO_2 storage capacity is 0.71, 0.67, 0.58, and 0.49 Mt respectively for the four types of injected gases, indicating that the effect of the temperature on the storage capacity decreases with the increasing impurity concentration in the injected gas. The solubility trapping capacity also decreases with the growing temperature. Particularly, it changes from 0.26 to 0.22 Mt for CO_2, and from 0.22 to 0.15 Mt for the gas with high concentration of impurity gas, which is result from the low solubility of CO_2 in high temperature condition.

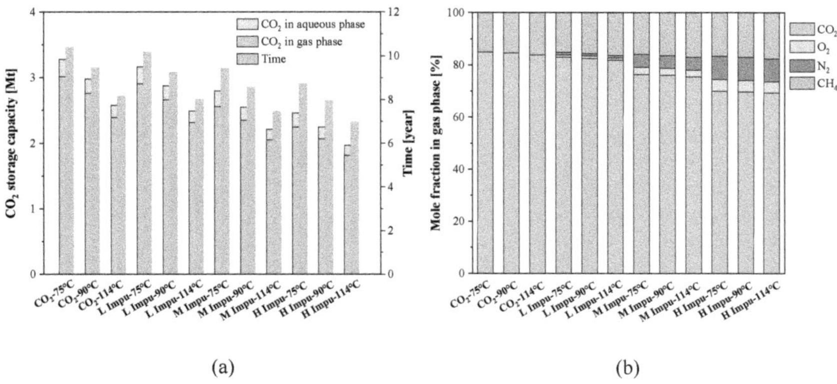

Figure 4.6 Main simulation results for the case of dedicated CO_2 storage with different reservoir temperature. (a) CO_2 storage capacity and the project duration; (b) Mole fraction of the gases in the reservoir

Fig. 4.6b shows that the effect of the temperature on the mole fraction of the gases in the gaseous phase is not significant due to the low solubility of impurities. However, it should be mentioned that the total gas in the reservoir with high temperature is less than that of low temperature. For example, the total gas corresponding the injected gas with high concentration of impurities are 59.5, 67.3, and 72.9 billion moles for the reservoir with temperatures of 114, 90, and 75 °C, respectively.

4.4.3 Effect of the residual water saturation

To investigate the effect of residual water saturation on the performance of CO_2 storage, the reservoir with an irreducible water saturation of 15% and 25% was investigated. In addition, considering that the flowable water often exists in the real gas reservoirs especially in the depleted gas reservoir associated with lateral or underlying aquifers, a reservoir with a residual water saturation of 55% (irreducible water saturation is 25%) was applied for simulation. The simulation results for the case of dedicated CO_2 storage with different water saturation are shown in Fig. 4.7.

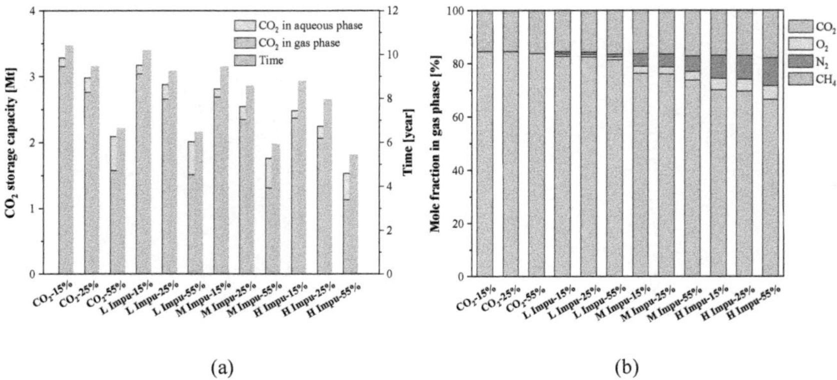

Figure 4.7 Main simulation results for the case of dedicated CO_2 storage with different water saturation. (a) CO_2 storage capacity and the project duration; (b) Mole fraction of the gases in the reservoir

As expected, it can be seen that both the storage capacity and the project duration decrease with the growing residual water. This is result from that less volume of pore is left for the injected gas related to the reservoir with more residual water. Actually, the situation of depleted gas reservoir is more similar with the deep saline aquifers with the increasing water saturation. Especially, when the water saturation in depleted gas reservoir increases up to 100%, the situation becomes to the same as the case of underground CO_2 storage in deep saline aquifers. Therefore, the decreasing CO_2 storage capacity with water saturation demonstrating the superiority of CCS in depleted gas reservoirs compared with saline aquifers.

The solubility trapping capacity increases from 0.12 to 0.2, and 0.45 Mt when the residual water saturation increases from 0.15 to 0.25, and 0.55 associated with the injected gas containing middle concentration of impurity gas. It can be calculated that the solubility trapping capacity is approximately proportional to the residual water saturation, this rule is also suitable for the other cases with different concentration of impurity gas. The reason account for this is that the injected CO_2 is abundant for dissolution compared with the residual water, thus the dissolution of CO_2 could reach to saturation under the associated pressure and temperature conditions. Due to the immobility of dissolved CO_2, it is considered as a safer trapping mechanism compared with the structural trapping in which CO_2 exits in free gaseous (Anchliya et al. 2012). Additionally, the dissolved CO_2 will increase the density of formation water by approximately 1%. Actually, such a small density difference is sufficient to promote the convection flow of the formation fluids, which is favorable for the solubility trapping of CO_2 (Cao et al. 2020; Zhang et al. 2008). In general, there is a contradiction between the total storage capacity and solubility trapping capacity, which should be balanced in the selection of depleted gas reservoirs for CO_2 sequestration.

Fig. 4.7b shows that the CO_2 mole fraction in gas phase decreases with the increasing residual water saturation. Specifically, it decreases from 84.6% to 83.8% for the pure CO_2, from 82.7% to 81.5%, from 76.4% to 73.9%, from 70.2% to 66.6% for the injected gases with low, middle, and high concentration of impurities, respectively. It can be calculated that a decrement of 0.8%, 1.2%, 2.5%, 3.6% was obtained for the four types of injected gases, demonstrating that the reduction of CO_2 impurity in the gas phases is more pronounced when the high concentration of impurities was co-injected.

4.4.4 Chromatographic partitioning phenomenon

The scheme of injecting high concentration of impurities with CO_2 into the reservoir with a residual water saturation of 0.25 and 0.55 respectively at the injection time of 0.5 year was regarded as examples to illustrate the mole fraction distribution of the gases in both gaseous and aqueous phases, which are shown in Fig. 4.8. It can be clearly seen that the swept area of the injected gases related to the reservoir with a residual water saturation of 0.55 is greater than that of 0.25. Fig. 4.8a shows that the mole fraction of CO_2, O_2, and N_2 in both gaseous and aqueous phase decreases monotonically away from the injection well for the reservoir with a residual water saturation of 0.25. Regarding the reservoir with a residual water saturation of 0.55, Fig. 4.8b shows that the mole fraction of CO_2 in gas phase decreases monotonically away from the injection well. While the mole fraction distribution of N_2 and O_2 in the gas phase are different from that of CO_2, which increase slowly away from the reservoir to reach a maximum value at the location of approximately 663 m away from the injection well, then decreases gradually to 0. This phenomenon also occurs in the case of injecting the gas with middle and low concentration of impurities as shown in Fig. 4.9.

Additionally, it should be mentioned that there is a high concentration of O_2 and N_2 in the aqueous phase around the injection well due to the constantly gas injection. Therefore, the concentrations of O_2 and N_2 in the aqueous phase decrease firstly, then increase to a maximum value and finally decrease again like the tendency in the gas phase (Fig. 4.10). Fig. 4.10 also shows that this phenomenon is more pronounced in the scheme with high concentration of impurity gases.

CO2 (g)
0.85
0.8
0.6
0.4
0.2
0
CO2 (aq)
0.0484
0.04
0.03
0.02
0.01
0
O2 (g)
0.05
0.04
0.02
0
O2 (aq)
1.94e-4
1.6e-4
1.2e-4
8.0e-5
4.0e-5
0
N2 (g)
0.0995
0.08
0.06
0.04
0.02
0
N2 (aq)
2.33e-4
2e-4
1e-4
0
CH4 (g)
0.99
0.8
0.6
0.4
0.2
0
CH4 (aq)
1.05e-3
1.0e-3
7.5e-4
5.0e-4
2.5e-4
0
0 200 400 600 800 1000 (m)

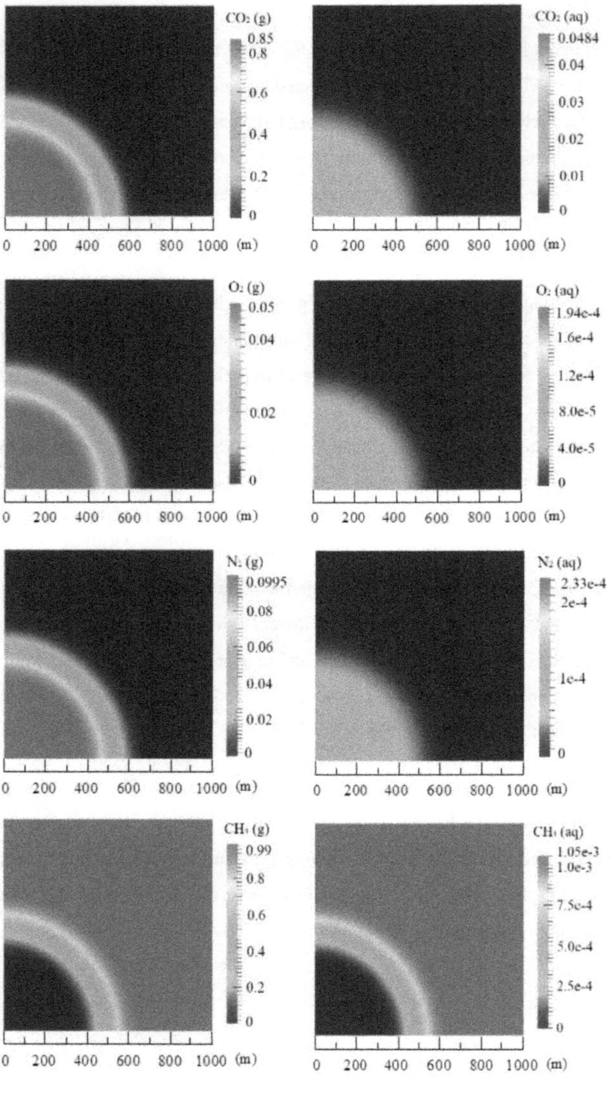

(a)

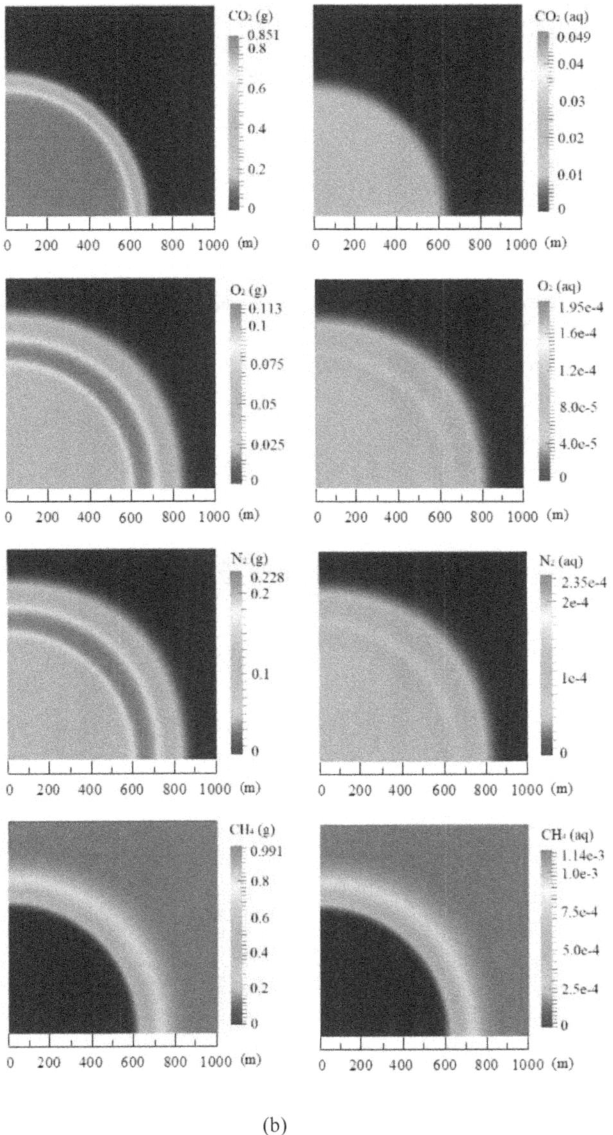

(b)

Figure 4.8 The mole fraction distribution of CO_2, O_2, N_2, and CH_4 in the gas and aqueous phases respectively associated with CO_2 injection with high concentration of impurities at the injection time of 0.5 year (ground plan of the reservoir). (a) The reservoir with a residual water saturation of 0.25 (only irreducible water distribution uniform in the depleted gas reservoir); (b) The reservoir with a residual water saturation of 0.55 (the flowable water increases from the top to the bottom of the reservoir)

CO2 (g)
0.851
0.8
0.6
0.4
0.2
0
CO2 (aq)
0.0526
0.04
0.02
0
O2 (g)
0.086
0.08
0.06
0.04
0.02
0
O2 (aq)
2.11e-4
2e-4
1e-4
0
N2 (g)
0.175
0.16
0.12
0.08
0.04
0
N2 (aq)
2.54e-4
2e-4
1e-4
0
CH4 (g)
0.995
0.8
0.6
0.4
0.2
0
CH4 (aq)
1.97e-3
1.6e-3
1.2e-3
8e-4
4e-4
0
0 200 400 600 800 1000 (m)

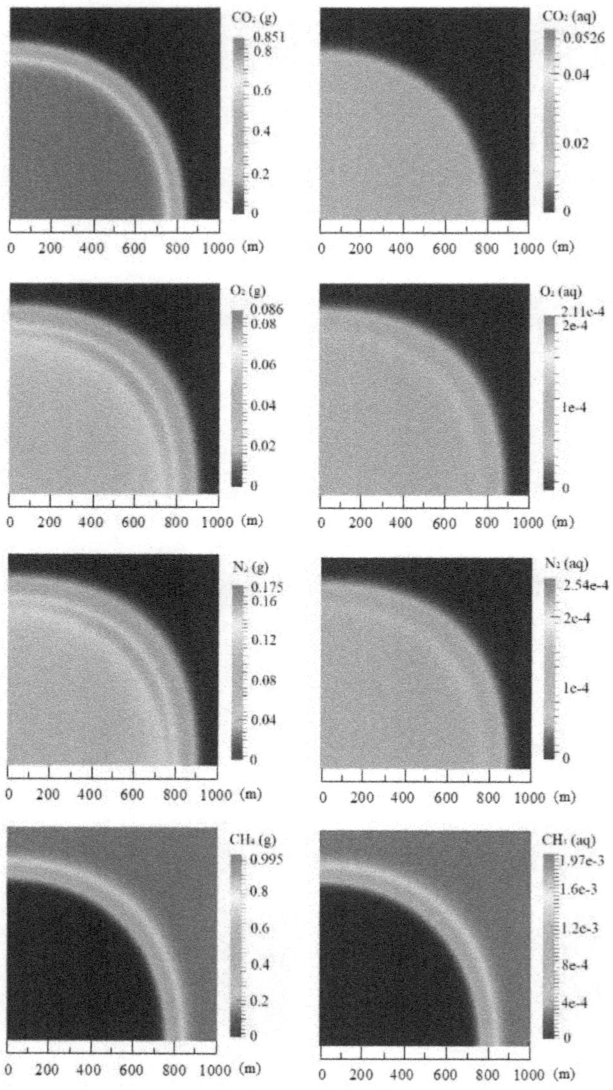

(a)

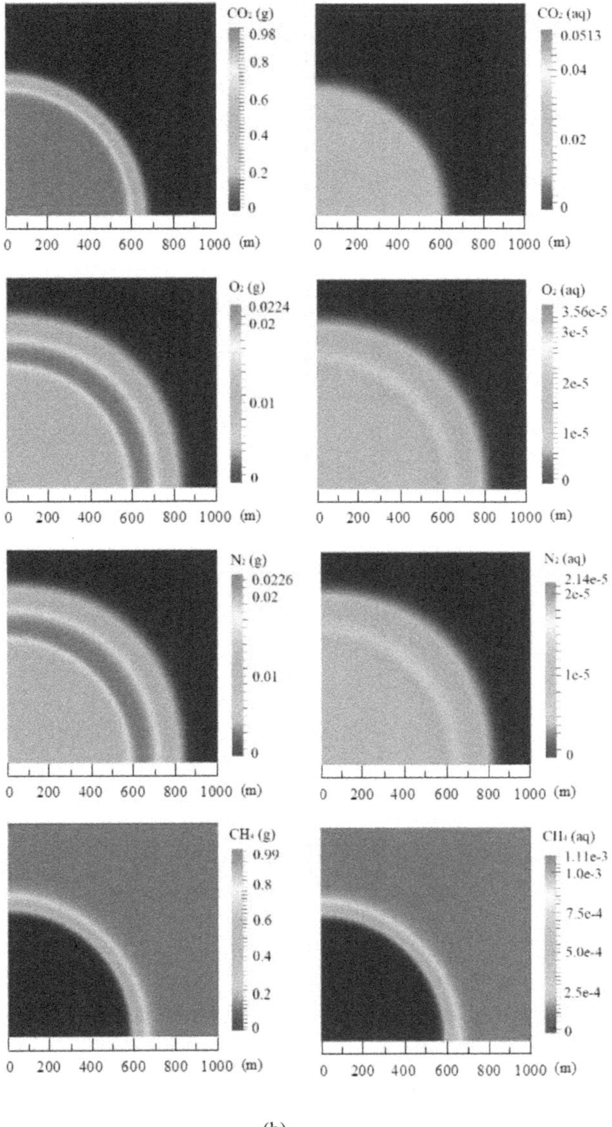

(b)

Figure 4.9 The mole fraction distribution of CO_2, O_2, N_2, and CH_4 in the gas and aqueous phases respectively for the reservoir with a residual water saturation of 0.55 at the injection time of 0.5 year with the injection of CO_2 with (a) middle and (b) low concentration of impurity gas (ground plan of the reservoir)

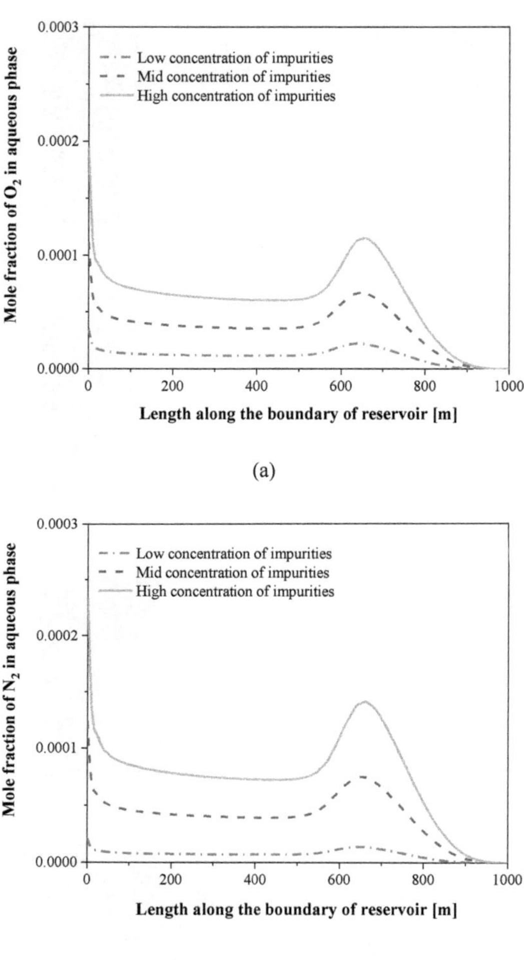

Figure 4.10 Mole fraction of (a) O_2 and (b) N_2 in the aqueous phase along the reservoir for the reservoir with a residual water saturation of 0.55

The forementioned mole fraction distribution difference of the gases (Fig. 4.8b and 4.9) is so called chromatographic partitioning phenomenon, which has been observed in both experimental studies (Bachu and Bennion 2009) and field test (Wei et al. 2015). The reason account for the chromatographic partitioning phenomenon is that CO_2 is more soluble than O_2 and N_2 at the reservoir temperature and pressure conditions. When the injected CO_2 with the impurity gases, i.e., O_2 and N_2 come in contact with the residual water in the depleted gas reservoirs, CO_2 dissolves preferentially until saturation is attained. In this way, the CO_2 could reduce the potential dissolution of O_2 and N_2 and keep them maintain a relatively high concentration in gas phase. As the injected gases migrate away from the injection well, they come in contact with unsaturated water which could dissolve the preferential CO_2 again. Therefore, the O_2 and N_2 migrate faster than CO_2 in the gas phase and concentrate at an area away from the injection well, which can be seen in Fig. 4.8b and 4.9. It should be pointed out that the chromatographic partitioning phenomenon also occurs in the aqueous phase result from the solubility difference of the injected gases (Fig. 4.8b and 4.9). This is consistent with the numerical results of injecting the mixture of CO_2 and impurities into the saline aquifers, which were conducted by Lei et al. (2016) and Wei et al. (2015).

The mole fraction of the gases in gaseous and aqueous phase respectively at the element in the diagonal of the reservoir subface and 920 m away from the injection well was record during the injection and illustrated in Fig. 4.11. The CO_2 migrates to this element in gas phase at the time of approximately 1 year, while the occurrence of O_2 is about 0.5 year and it occurs earlier with the higher concentration of impurities. Fig. 4.11c, 11e, and 11g show that the occurrence of N_2 is slightly earlier compared with O_2, which is result from its lower solubility at the associated pressure and temperature conditions. This situation also occurs in the aqueous phase (Fig. 4.11d, 11f, 11h) and this phenomenon is more pronounced in the scheme with high concentration of impurity gases, which is in accordance with Li and Jiang (2017)'s results obtained by co-injection of N_2 and CO_2 into deep saline aquifers. It means that the N_2 and O_2 can be detected approximately 0.5 year before the CO_2 at the determined location, which can play an important role in the monitoring procedure. For instance, the detection of N_2 and O_2 front could be regarded as a signal of CO_2 front and potential CO_2 leakage after a time lag.

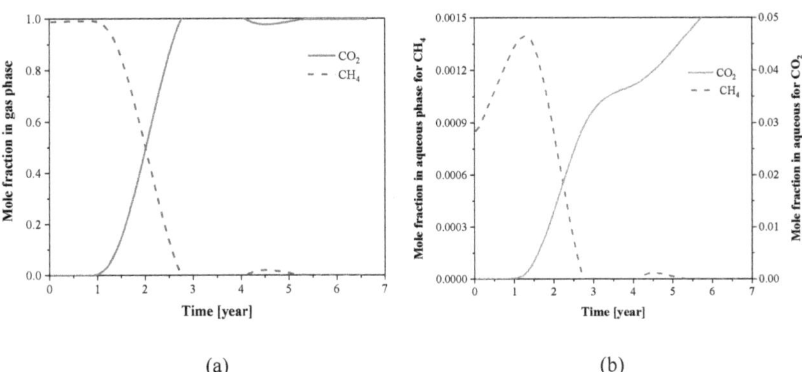

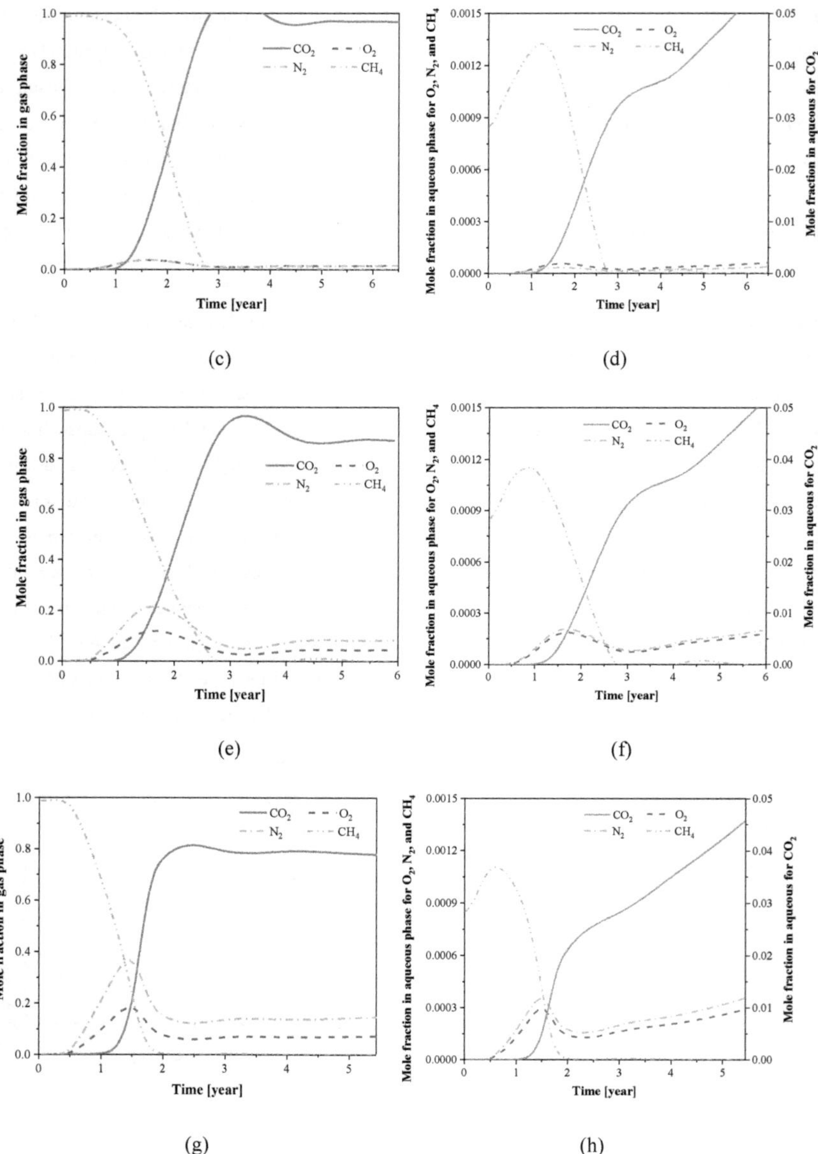

Figure 4.11 The mole fraction of the gases in gaseous and aqueous phase at a determined point for variety of injected gases. (a-b) The scheme of pure CO_2 injection; (c-d) CO_2 injection with low concentration of impurities; (e-f) CO_2 injection with middle concentration of impurities; (g-h) CO_2 injection with high concentration of impurities

It should be mentioned that there is no apparent chromatographic partitioning phenomenon at the top surface of the reservoir with a residual water saturation of 0.55, as shown in Fig. 4.12. The main reason account for this is the relative lower water saturation at the top layer due to the flowing downward of flowable water dominated by gravity. As shown in Fig. 4.13, the water saturation in the reservoir increases from approximately 0.46 to 0.85 from the top layer to bottom layer. Another reason account for the absence of the chromatographic partitioning phenomenon at the top layer is the relatively lower concentration of injected gases in this zone due to the relatively lower vertical permeability compared with the horizontal permeability. Therefore, it is more efficient to detect the N_2 and O_2 front at the bottom layer of reservoir and take it as a signal of CO_2 front and potential CO_2 leakage.

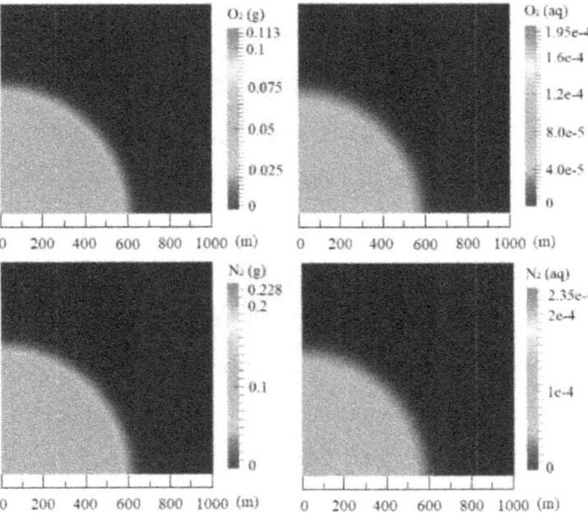

Figure 4.12 The mole fraction distribution of O_2 and N_2 in the gas and aqueous phases respectively at the injection time of 0.5 year associated with CO_2 injection with high concentration of impurities into the reservoir with a residual water saturation of 0.55 (top view of the reservoir)

Figure 4.13 The distribution of water saturation for the reservoir with a residual water saturation of 0.55 (gradually increase from approximately 0.46 to 0.85 from the top layer to bottom layer result from the flowing downward of flowable water dominated by gravity)

Bases on the forementioned analysis, it can be concluded that the difference of solubility among the injected gases dominates the chromatographic partitioning phenomenon, and the concentration of the impurity gases is considered a secondary factor affecting it, which is consistent with the results of Bachu et al. (2009). Additionally, the water saturation in the reservoir is a key factor for the chromatographic partitioning phenomenon. It can be inferred that there is a critical water saturation for the occurrence of significant chromatographic partitioning phenomenon associated with determined type and concentration of impurity gas.

It can be seen from Fig. 4.8b and 4.9 that the CH_4 concentrated at the area away from the injection well and maintained a relatively high purity, indicating the potential of CH_4 production enhanced by the injection of CO_2 with impurities, which will be discussed in the following section.

4.5 CSEGR

4.5.1 Effect of the injection rate

Fig. 4.14 shows the main simulation results for CSEGR with different injection rate, i.e., 2.5, 5, and 10 kg/s, based on the depleted gas reservoir with a primary gas recovery of 80%. It can be seen from Fig. 4.14a that the CO_2 storage capacity increases during EGR whereas decreases during CCS with the increasing of injection rate for the four types of injected gases. For a certain injection rate, the CO_2 storage capacity during EGR decreases with the growing impurity concentration. For example, the storage CO_2 capacity during EGR with the injection rate of 5 kg/s are 1.28, 1.24, 1.1, and 1.01 Mt for CO_2, and the gas with low, middle, and high concentration of impurities, respectively. Fig. 4.14a also shows that the stored CO_2 in gas phase and aqueous phase is almost identical for a certain type of injected gas. The effect of injection rate on the total CO_2 storage capacity is neglectable, no matter what composition of the injected gas is. Fig. 4.14b shows that the effect of injection rate on the mole fraction of the gases in the reservoir when the projects are terminated is minor. For example, regarding the injection of CO_2 with high concentration of impurity gas, the mole fraction of CO_2 varies from 76.45% to 75.69% when the injection rate varies from 2.5 to 10 kg/s. The corresponding changes of the mole fraction of O_2, N_2, and CH_4 are from 4.86% to 4.81%, from 9.69% to 9.59%, and from 9% to 9.91%, respectively. This means that the compositions in the reservoir are almost identical except for CH_4. In this scheme, the EGR varies from 9.19% to 8.12% as shown in Fig. 4.14c. It shows that the EGR decreases with the growing injection rate. Particularly, the EGR decreases by approximately 1% when the injection rate changes from 2.5 to 10 kg/s. It can be seen from Fig. 4.14c that the project duration decreases dramatically with the rising of injection rate, especially the time for EGR, which varies from 12.5 to 5.3 years for CO_2 and from 11.2 to 4.6 years for the injected gas with high concentration of impurities, demonstrating the significant impact of the injection rate on the mixing of the gases in the reservoir. Because the operation of EGR is terminated when the concentration of CH_4 in the produced gas is lower than 90%, less project duration during EGR means high extent of mixing for the gases. The project duration corresponding the low injection rate (2.5 kg/s) for the four types of gases are 20.1, 19.7, 18.4, and 17.4 year, respectively. While the vale corresponding to high injection rate (10 kg/s) are 10.5, 10.3, 9.5, and 8.8 year respectively for the four types of gases. Considering that the total CO_2 storage capacity is almost identical (Fig. 4.14a), it can be concluded that only around 1% of incremental EGR could be achieved at the additional cost of about 9 years for the injection rate of 2.5 kg/s compared with 10 kg/s, demonstrating the superiority of high injection rate in the scheme of CSEGR.

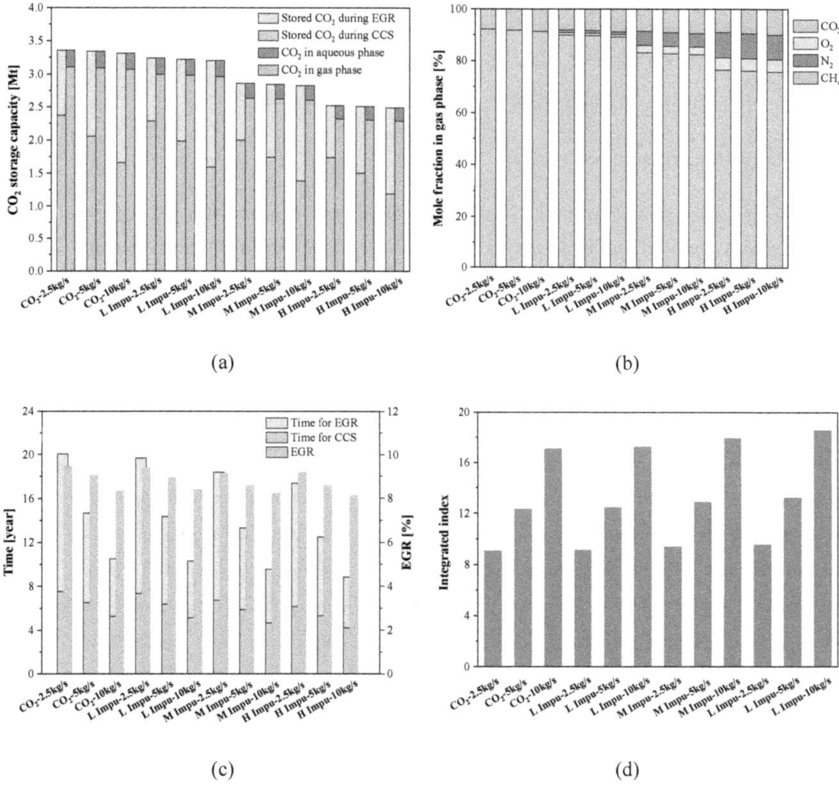

Figure 4.14 Main simulation results for CSEGR with different injection rate. (a) CO_2 storage capacity; (b) Mole fraction of the gases in the reservoir; (c) Project duration and the EGR; (d) The integrated index

The mole fraction of CO_2 in the reservoir is an indicator to characterize the usage efficiency of the reservoir for CO_2 storage. The value of EGR and project duration can directly characterize the performance on CH_4 production and the time costs for the projects. Hence, these three factors, i.e., the EGR performance, the mole fraction of CO_2 in the reservoir, and the total time spent on the project could be regarded as the most important factors for CSEGR at a depleted gas reservoir. Therefore, an integrated index is defined to compare the performance of CO_2 storage associated with EGR for the kinds of schemes. The integrated index is calculated using the sum of CO_2 fraction (Fig. 4.14b, %) and

EGR (Fig. 4.14c, %) divided by the project duration (Fig. 4.14c, year) as shown in Fig. 4.14d. Compared with the evaluation index proposed by Zhang et al. (2017), the usage efficiency of reservoir for CO_2 storage was considered for the integrated index in this work, thus it can evaluate the comprehensive performance of CSEGR better. Fig. 4.14d shows that the schemes with an injection rate of 10 kg/s has the same and the largest value of integrated index, demonstrating the superiority of high injection rate in the scheme of CSEGR again. Therefore, a high injection rate is recommended to be used for CSEGR. In the following studies, the higher injection rate (10 kg/s) will be used.

4.5.2 Effect of the primary gas recovery

Fig. 4.15 shows the simulation results for the purpose of CSEGR at depleted gas reservoirs with a primary gas recovery of 70%, 80%, and 90%, respectively. Obviously, a higher CO_2 storage capacity could be attained in a reservoir with high gas recovery. It can be seen from Fig. 4.15a that the CO_2 storage capacity increases from 3.12 to 3.53 Mt for CO_2, from 3 to 3.4 Mt, 2.66 to 3.01 Mt, 2.34 to 2.65 Mt for the injected gas with low, middle, and high concentration of impurities when the primary gas recovery increases from 70% to 90%. The solubility trapped CO_2 slightly increases with the primary gas recovery. Particularly, it changes from 0.23 to 0.26 Mt for CO_2, and from 0.19 to 0.21 Mt for the gas with high concentration of impurities. The trapped CO_2 decreases during EGR whereas it increases during CCS with the increasing primary gas recovery. This is result from that more impurity gas is likely to be produced for the reservoir with a high primary gas recovery, thus the corresponding project duration for EGR is relatively short. As shown in Fig. 4.15c, the time for EGR varies from 6.2 to 3.89 years for CO_2, and from 6.09 to 3.82 years, from 5.74 to 3.64 years, from 5.41 to 3.42 years for the gases with low, middle, and high concentration of impurities. Of course, more time is needed to recovery to the initial pressure of reservoir for the schemes with high primary gas recovery (Fig. 4.15c). Fig. 4.15c also depicts that the performance of EGR is highly dependent on the primary gas recovery of the reservoir, while the impact of impurity gas is minor. Specifically, the EGR are approximately 13.6%, 8.3%, and 3.3% for the reservoirs with a primary gas recovery of 70%, 80%, and 90%, respectively. It can be calculated that the overall gas recoveries are 83.6%, 88.3%, and 93.3% for the reservoirs with three different primary gas recovery. This is consistent with Jikich et al. (2003)'s results, which reveals that CO_2 injection after the gas reservoir becomes depleted is the best scenario for maximizing the total gas production, whereas leads to a low incremental gas recovery. Fig. 4.15d shows that the integrated index is almost the same for a certain type of injected gas. Considering that a high CO_2 storage capacity and high CO_2 concentration can also be attained for the scheme of high primary gas recovery (Fig. 4.15a and 15b), it is suggested to produce the CH_4 as possible before the operation of CO_2-EGR.

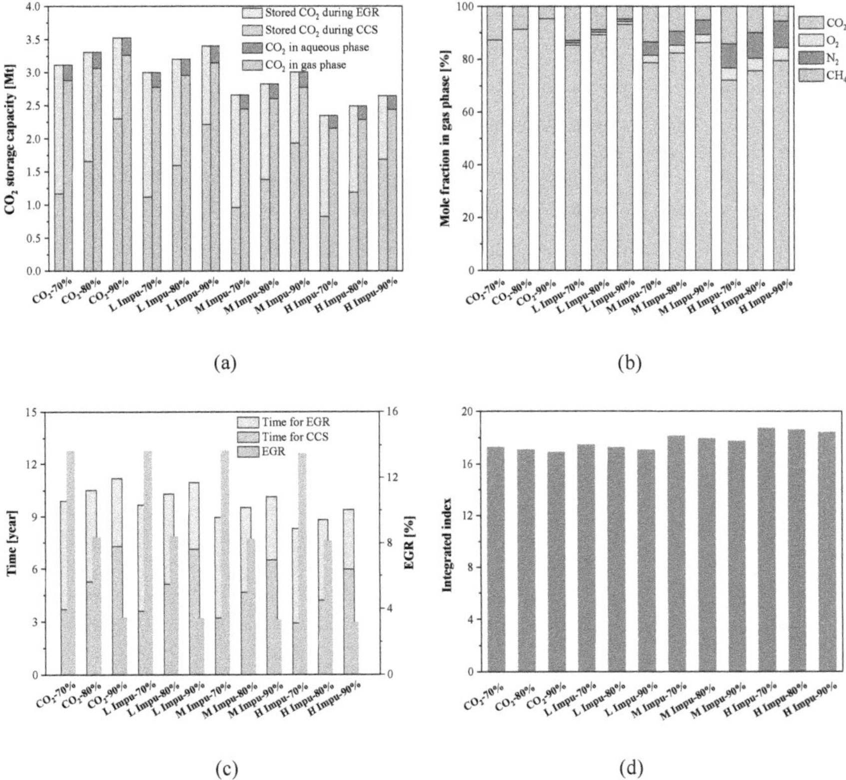

Figure 4.15 Main simulation results for CSEGR with different primary gas recovery. (a) CO_2 storage capacity; (b) Mole fraction of the gases in the reservoir; (c) Project duration and the EGR; (d) The integrated index

4.5.3 Effect of the reservoir temperature

Fig. 4.16 depicts the simulation results for CSEGR with a reservoir temperature of 75, 90, and 114 °C, respectively. It can be seen from Fig. 4.16a that the CO_2 storage capacity decreases with the growing reservoir temperature. Specifically, the CO_2 storage capacity varies from 3.64 to 2.86 Mt for CO_2, and from 3.53 to 2.77 Mt, from 3.11 to 2.45 Mt, from 2.74 to 2.16 Mt for the gas with low, middle, and high concentration of impurities. It can be calculated that the decrement of CO_2 capacity is 0.78, 0.76, 0.66, and 0.58 Mt respectively for the four types of injected gases, indicating that the effect of the reservoir temperature is decreasing with the increasing impurity concentration in the injected gas. The solubility trapping capacity also decrease with the increases of temperature due to its impact on the solubility of CO_2. Fig. 4.16b demonstrates that the effect of the temperature on the mole fraction of the gases in the gaseous phase is minor. Fig. 4.16c illustrates that the project duration decreases with the increasing impurity concentration and the temperature account for the low compressibility of N_2 and O_2 and the high pressure caused by high temperature. Actually, the forementioned analysis is the same as the scenario for the purpose of dedicated CO_2 storage.

Fig. 4.16c also shows that the impact of reservoir temperature on the EGR is not significant. The reservoir with high temperature has a relatively large integrated index (Fig. 4.16d), indicating a high usage efficiency could be achieved in a reservoir with high temperature condition. However, it should be pointed out that only a relatively low storage capacity could be attained account for the high reservoir pressure caused by high temperature in this case (Fig. 4.16a).

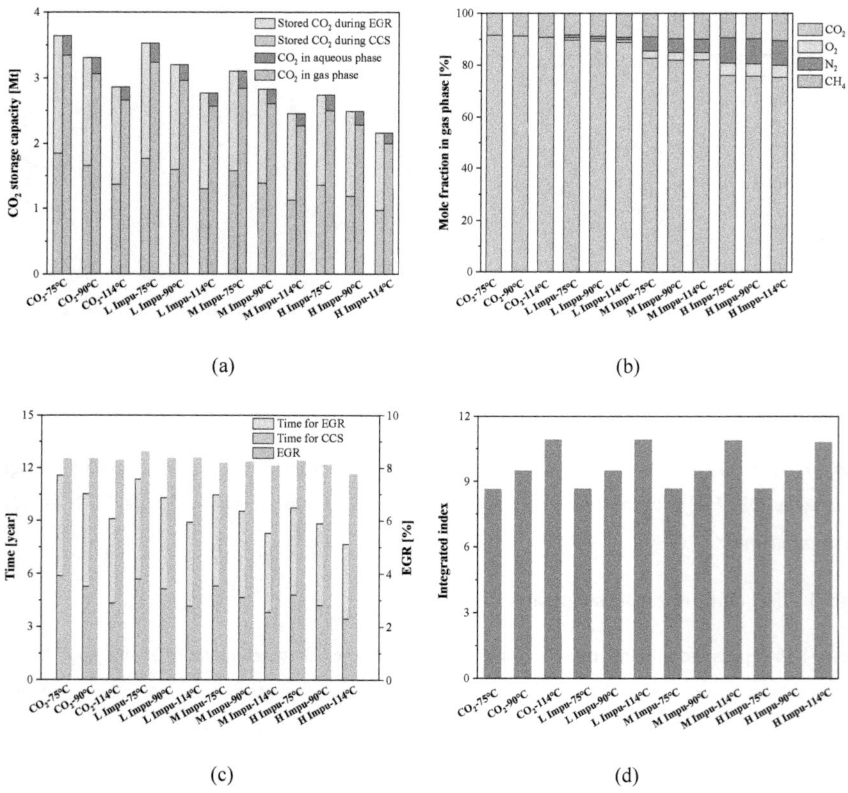

Figure 4.16 Main simulation results for CSEGR with different reservoir temperature. (a) CO_2 storage capacity; (b) Mole fraction of the gases in the reservoir; (c) Project duration and the EGR; (d) The integrated index

4.5.4 Effect of the residual water saturation

Considering that a lot of water would be produced when the flowable water exists in the depleting gas reservoirs which is not favorable for EGR, thus the reservoirs with only irreducible water were studied in this thesis. The main simulation results for CSEGR with irreducible water saturation of 0.15, 0.25, and 0.35 respectively was depicted in Fig. 4.17. It can be seen that the CO_2 storage capacity decreases significantly with the increasing of water saturation. Specifically, it decreases from 3.64 to 2.99 Mt for CO_2 when the water saturation increases from 0.15 to 0.35, and decreases from 3.52 to 2.88 Mt, from 3.12 to 2.55 Mt, from 2.74 to 2.23 Mt for the injected gas with low, middle, and high concentration of impurities, respectively. The decrement of CO_2 storage capacity declines with the growing of impurity gas concentration. Whereas the solubility trapping capacity increases from 0.15 to 0.34 Mt for CO_2, and from 0.12 to 0.28 Mt for the injected gas with high concentration of impurities. The CO_2 concentration in the gas phases is almost identical for a certain type of injected gas as shown in Fig. 4.17b. Fig. 4.17c demonstrates that the project duration decreases with the increasing water saturation account for less volume of pores remained. Actually, the above analysis is similar with the situation of the scheme for the purpose of dedicated for CO_2 storage.

Fig. 4.17c also shows that the project duration for EGR decreases with the rising of residual water saturation, indicating the gas is more mixed for the reservoir with high water saturation. This is result from that the operation of EGR is terminated when the concentration of CH_4 is lower than 90%. This could be explained by the following reasons. The first one is that the water occupied pores promote the injected gas traveling away from the injection well and mixing with the CH_4. The second one is that the water occupies pores and creates more tortuous flow-paths for the gases in the reservoir, which has been identified by the analysis of nuclear magnetic resonance during a core flood experiment (Honari et al. 2016). Another reason is that the total volume of CH_4 is relatively less for the reservoir with high irreducible water saturation under the same primary gas recovery. For example, the initial mass of CH_4 in depleted gas reservoir with an irreducible water saturation of 0.15 and 0.35 are 2.09×10^8 and 1.61×10^8 kg respectively, thus the displacement effect is more pronounced in the case of high residual water saturation with the same injection rate. It also caused the relatively high EGR for the scheme with high irreducible gas saturation (Fig. 4.17c). The reservoir with high irreducible water saturation has a relatively large integrated index (Fig. 4.17d), indicating a high usage efficiency could be achieved for this scheme. However, it should be pointed out that only a relatively low storage capacity could be attained account for the less efficient pores remained caused by the non-compressible water in this case (Fig. 4.17a).

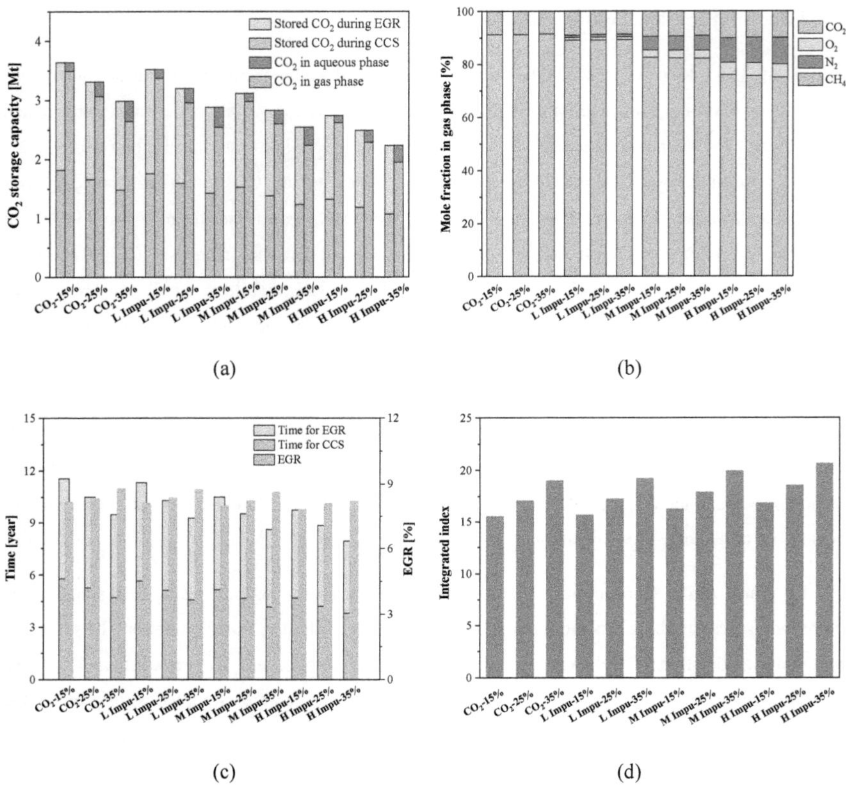

Figure 4.17 Main simulation results for CSEGR with different residual water saturation. (a) CO_2 storage capacity; (b) Mole fraction of the gases in the reservoir; (c) Project duration and the EGR; (d) The integrated index

4.5.5 Effect of the threshold impurity concentration in EGR

Fig. 4.18 illustrates the main simulation results for CSEGR with different threshold impurity concentration during EGR. In other words, the operation of EGR is terminated when the impurities in the produced gas reaches to different threshold concentration, i.e., 10%, 20%, and 30%, respectively. As expected, it can be seen from Fig. 4.18a that the CO_2 storage capacity increases with the growing threshold impurity concentration during the process of EGR. For the case of pure CO_2 injection, the CO_2 storage capacity increases from 3.31 to 3.37 Mt when the threshold impurity concentration in EGR increases from 10% to 20%. It can be calculated that only an increment of CO_2 storage capacity of 0.06 Mt was obtained, while the increment of CO_2 storage capacity decreases to 0.04 Mt when the threshold impurity concentration in EGR increases from 20% to 30%, demonstrating the decreasing effect of threshold impurity concentration on the CO_2 storage capacity. This tendency also occurs in the case of CO_2 injection with impurity gas. Fig. 4.18a also shows that the stored CO_2 in aqueous phase is almost identical for a certain type of injected gas. Fig. 4.18b shows that the mole fraction of CO_2 in the gas phase increases by approximately 1.2% when the threshold impurity concentration in EGR increases from 10% to 20%, while the increment is only approximately 0.8% when the threshold impurity concentration increases from 20% to 30% due to more CO_2 was produced. This can also be demonstrated by the project duration in Fig. 4.18c. It demonstrates that an additional time of approximately 0.5 year is needed when the threshold impurity concentration increases from 10% to 20%, while it decreases to approximately 0.4 year when the threshold impurity concentration increases from 20% to 30%. The EGR in these two stages are approximately 1.5% and 1%, respectively (Fig. 4.18c). This is result from that the production rate of impurity gas increases gradually during the EGR process. In other words, more impurity gas was produced together with CH_4, which is not favorable for the overall benefits because it increases the cost on gas separation. Fig. 4.18c also shows that the effect of N_2 and O_2 on the EGR is not significant, while less time is needed when they are co-injected because of their lower compressibility compared with that of CO_2. Therefore, the scenario with high concentration of impurity gas has a relatively large integrated index (Fig. 4.18d). However, it should be pointed out that only a relatively low storage capacity could be attained account for the low compressibility in this case (Fig. 4.18a).

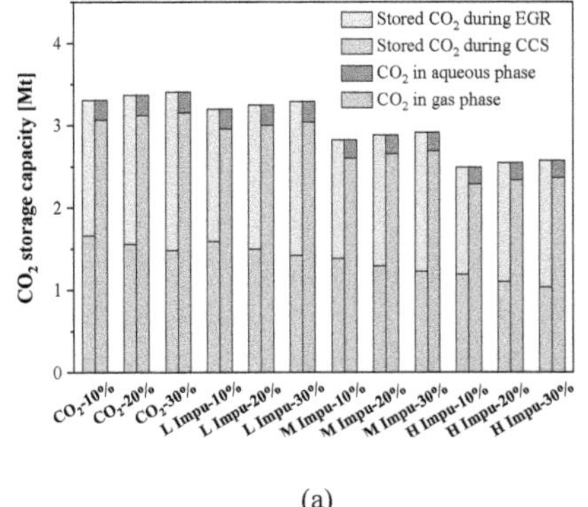

(a)

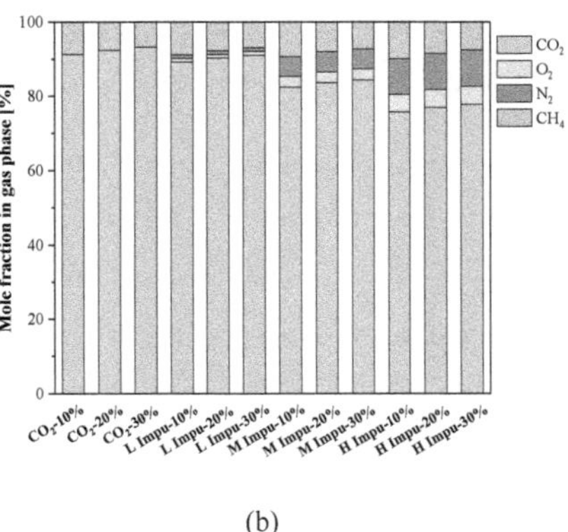

(b)

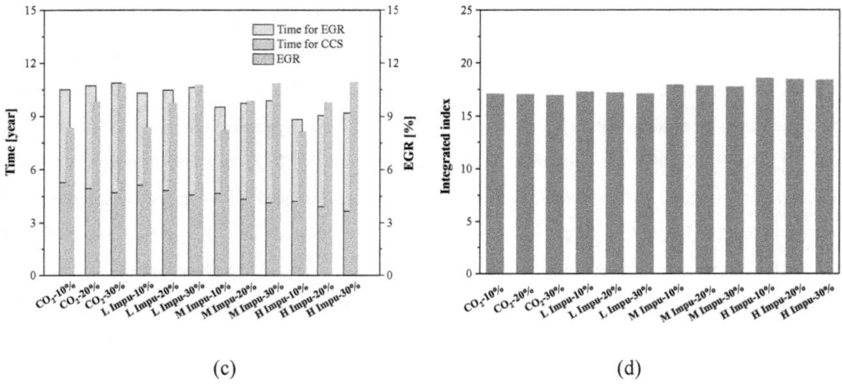

Figure 4.18 Main simulation results for CSEGR with different threshold impurity concentration during EGR. (a) CO_2 storage capacity; (b) Mole fraction of the gases in the reservoir; (c) Project duration and the EGR; (d) The integrated index

4.6 Preliminary economic analysis

Based on the related economic parameters as shown in Tab. 4.4 (Irlam 2017; Shen et al. 2014; Tseng et al. 2011), the economic effects of injecting pure CO_2 into depleted gas reservoir with a primary gas recovery of 70%, 80% and 90% were analyzed. The fiscal subsidies balanced prices for CCS and CSEGR with CO_2 injection are shown in Fig. 4.19.

Table 4.4 Main parameters of the reservoir simulation model (Irlam 2017; Shen et al. 2014; Tseng et al. 2011)

Well cost CO_2 injection well (US $ Million/well)		Operating cost			
		Fixed cost (US $ Million/year)		Variable cost (US $ /ton)	
Repair payment	Installment payment	CO_2 tank hired	Monitoring cost	CO_2	Operating cost
0.033	0.033	0.06	0.133	60	0.067
Gas price and operating cost			Tax and others		
Natural gas (US $ SCM)	Fixed cost (US $ Million/year)	Variable cost (US $ /SCM)	Royalties (US $ /SCM)	Business tax (%)	
0.42	0.067	0.073	0.026	25	

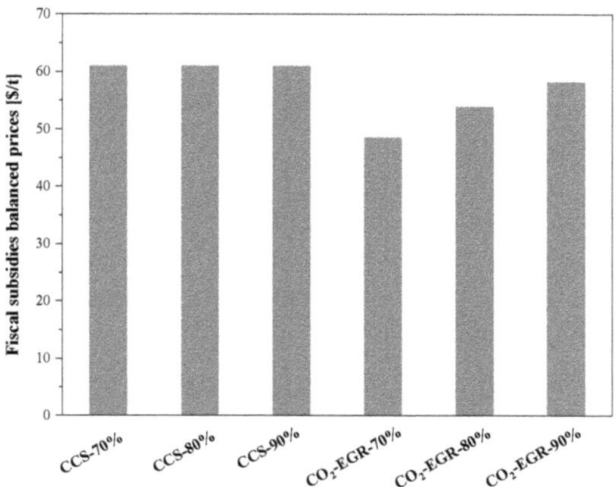

Figure 4.19 Fiscal subsidies balanced prices for CCS and CSEGR with CO_2 injection

It can be seen that the fiscal subsidies balanced prices for the case of dedicated CO_2 storage are almost identical and slightly higher than the cost of CO_2, demonstrating that the capture and separation of CO_2 dominate the overall cost of CCS. This is in accordance with the previous results (GCCSI 2011; Gislason and Oelkers 2014). It also indicates that co-injection of CO_2 with impurity gas would reduce the overall cost of CCS by the reduction of the cost on gas separation. It should be mentioned that the effect of impurity gas on the cost of CCS has not been analyzed in this work because of the lack of economic data related to the effect of impurities on the cost of separation and transportation, which would be analyzed in future study.

Regarding the case of CSEGR, the fiscal subsidies balanced prices are 48.51, 53.86, and 58.23 dollars/ ton for the scenario with a primary gas recovery of 70%, 80%, and 90%, respectively. This means that the economic benefit of CSEGR is not significant in the reservoir with a high primary gas recovery. Overall, it is suggested to apply CSEGR in the depleted gas reservoir with a low primary gas recovery, while apply dedicated CO_2 storage in the case of high primary gas recovery.

4.7 Summary

The CO_2 contained the impurities of N_2 and O_2 is feasible to be injected into depleted gas reservoirs for CO_2 storage and CSEGR. Generally, the impact of impurities on the CO_2 storage capacity is dependent on the reservoir pressure and temperature conditions, and the concentration of the impurities. The depleted gas reservoir with a relatively low temperature and low irreducible water saturation is favorable for the CO_2 storage capacity. In the given depleted gas reservoir in this work, the CO_2 storage capacity decreases proportionally to the concentration of impurities.

It is suggested to produce the CH_4 as possible before the operation of CO_2 storage and CSEGR. The effect of injection rate on the total CO_2 storage capacity in the scenario of CSEGR is neglectable, whereas a high injection rate is beneficial for CSEGR due to the corresponding short project duration. The depleted gas reservoir with a low primary gas recovery is favorable for CSEGR, while it is suitable for dedicated CO_2 storage for the reservoir with a high primary gas recovery.

The chromatographic partitioning phenomenon may occur in both gaseous and aqueous phases when the N_2 and O_2 are co-injected into the depleted gas reservoirs, which could be used as a monitoring strategy for the CO_2 front and potential CO_2 leakage. In addition to the solubility and concentration of the impurity gas would affect this phenomenon, there is a critical water saturation for the occurrence of significant chromatographic partitioning phenomenon associated with determined type and concentration of impurity gas.

5 CO_2 as cushion gas for underground natural gas storage in depleted gas reservoirs

For the purpose of investigating the suitability of utilizing CO_2 as the cushion gas in the underground gas storage reservoir (UGSR), the Donghae depleted gas reservoir located in Ulleung basin in Korea is used. The CO_2 was firstly injected into the USGR as cushion gas, then the cyclic CH_4 production and injection was performed over a period of 15 years, which reaches a real engineering operation and makes it more beneficial for characterizing the mixing behavior of CO_2 and CH_4 in a relatively long-term period. Additionally, the impact of the CO_2 fraction, reservoir temperature, reservoir permeability, residual water, and production rate on the reservoir average pressure, the mixing behavior, and CO_2 concentration in the produced gas were investigated respectively in detail. The main contents of this chapter have been published in the following research paper (Cao et al., 2020): Utilization of CO_2 as cushion gas for depleted gas reservoir transformed gas storage reservoir. *Energies*, *13*(3), 576.

5.1 Simulation schemes

5.1.1 Geological model based on the Donghae gas reservoir in Korea

A simple brick model with the dimension of 914.4 m × 914.4 m × 30.48 m is used based on the geological information of the Donghae gas reservoir from the Ulleung basin in Korea, in which a slightly anticline structure is neglected (Kim et al. 2015). As depicted in Fig. 5.1a, a typical five-spot pattern was applied in this study, wherein one well is located at the lower layer for CO_2 injection, and the other four wells are located at upper layer for the CH_4 injection and production. Due to the symmetry of this reservoirs, only a quarter model of the reservoir is selected for the computational simulation (Fig. 5.1b). The porosity of the reservoir is 0.2, and the horizontal and vertical permeability are 50 and 10 mD, respectively. The depleted pressure of the reservoir is 5.17 MPa, with a gas saturation of 80%. The geological properties of the reservoir are exhibited in Tab. 5.1.

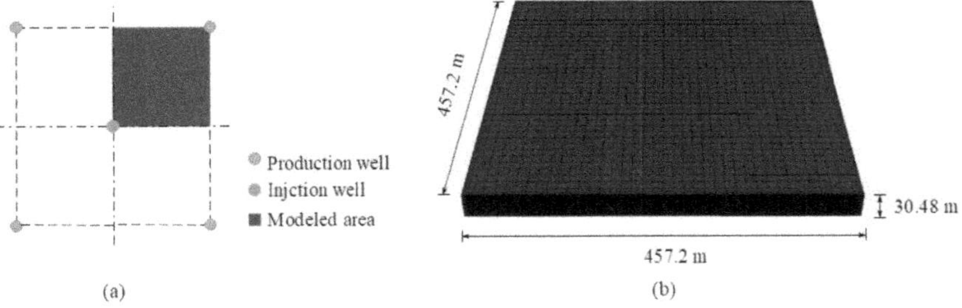

Figure 5.1 (a) Five-spot pattern depicting the CO_2 injection well and the production wells; (b) A quarter model of the reservoir in the Ulleung basin. (Cao et al. 2020; Kim et al. 2015)

Table 5.1 Reservoir parameters (Kim et al. 2015)

Parameters	Value
Porosity	0.2
Horizontal permeability [mD]	50
Vertical permeability [mD]	10
Gas saturation [-]	80%
Initial pressure [MPa]	5.17

5.1.2 Parameters for numerical simulations

There are three components in the multiphase flow of the UGSR, including CO_2, CH_4, and water. During the process of simulation, the diffusion coefficient is a vital parameter affecting the mixing behavior of CO_2 and CH_4 in the UGSR. In this work, an empirical model was used to characterize the pressure and temperature dependent diffusion coefficient, which can be written as follows (Pruess et al. 1999; Walker et al.1981):

$$d_\beta^k(P,T) = d_\beta^k(P_0,T_0)\frac{P_0}{P}\left(\frac{T+273.15}{273.15}\right)^\theta \tag{5.1}$$

where P_0 and T_0 represents the standard conditions, which equals 0.1 MPa and 0 °C, respectively. The diffusion coefficient at this condition for CO_2 and CH_4 has a value of approximately 10^{-5} m^2/s. The temperature dependence parameter θ is 1.8. This equation is an intrinsic equation integrated in the original version of TMVOC EOS/TOUGH2MP, in which it can be applied under the temperature and pressure ranging 0~350 °C and 0.1~100 MPa, respectively.

To verify the validity and suitability of Eq. 5.1, the calculated diffusion coefficients of CO_2-CH_4 are compared with the experimental results from Honari et al. (2015). In their work, the diffusion coefficients of CO_2-CH_4 for the Estaillades carbonate, Donnybrook sandstone, and Ketton carbonate were measured by a core flooding apparatus as shown in Fig. 5.2.

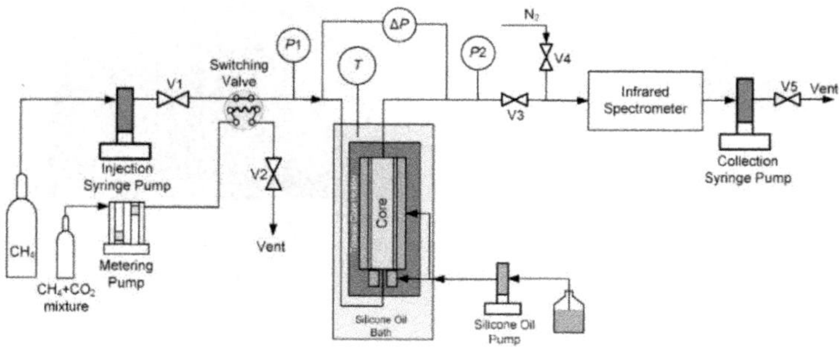

Figure 5.2 Schematic of the core flooding apparatus (Honari et al. 2015)

The diffusion coefficients of CO_2-CH_4 obtained from the experiments and the empirical function are summarized in Tab. 5.2.

Table 5.2 The comparation of the diffusion coefficients of CO_2-CH_4 systems obtained from experimental and empirical function

Core	T [°C]	P [MPa]	Diffusion coefficient [10^{-8} m^2/s]		Relative error
			Experimental results from Honari et al. (2015)	Calculated results by Eq. 5.1	
Estaillades carbonate	40	8	16.4	15.99	2.52%
	40	10	12.5	12.79	2.31%
	40	12	10	10.66	6.57%
	60	10	15.2	14.30	5.94%
	60	12	12.3	11.91	3.14%
Donnybrook sandstone	40	12	10	10.66	6.57%
	40	14	8.1	9.13	12.78%
	60	10	15.2	14.30	5.94%
	60	12	12.3	11.91	3.14%
Ketton carbonate	40	8	16.4	15.99	2.52%
	40	10	12.5	12.79	2.31%
	40	12	10	10.66	6.57%
	40	14	8.1	9.13	12.78%
	60	10	15.2	14.30	5.94%

As shown in Tab. 5.2, the calculated diffusion coefficients by Eq. 5.1 are approach to the experimental results for all the three kinds of cores. Specifically, the error percentage is between 2.31% and 12.78%, and most of them are lower than 7%. Therefore, it can be concluded that the Eq. 5.1 can be used for characterizing the diffusion behaviour of CO_2 and CH_4 in the UGSR. The empirical model has been implemented into the multicomponent and multiphase flow simulator TOUGH2MP that is used in this work.

5.1.3 Operation scenarios

Considering that the depleted pressure of the reservoir is low, two years of gas injection, i.e., CO_2 injection in the first year and CH_4 injection in the second year, is implemented before its operation as a gas storage reservoir. The purpose of injecting gas is to recover the average reservoir pressure from 5.17 MPa to the original reservoir pressure of 24 MPa (Lee et al. 2010). Thereafter, the UGSR was in operation for 15 years. The CH_4 is produced in winter and injected in summer in this UGSR, with the purpose of meeting the widely fluctuating demand of natural gas. As depicted in Fig. 5.3, the UGSR works in one-year cycles and each cycle consists of four stages. In the first stage, CH_4 is produced at a rate of 4.05 kg/s from Nov. 1st to Feb. 28th the next year (120 days), followed by well shutting and facility checking from Mar. 1st to Apr. 4th (35 days). In the third stage, CH_4 is injected at a rate of 2.7 kg/s from Apr. 5th to Oct. 1st (180 days). Then the well is shut again from Oct. 2nd to Oct. 31st (30 days).

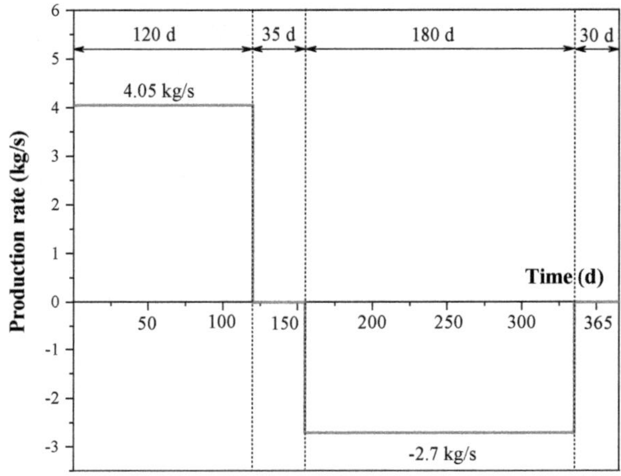

Figure 5.3 Production and injection rate in a one-year cycle (Cao et al. 2020)

5.2 Properties and mixing behavior of the gases in the UGSR

5.2.1 Physical properties of the mixed gases

The mixing behavior of CO_2 and CH_4 in a UGSR is determined by many factors, including the density differences, mobility ratios, molecular diffusion, and mechanical dispersion (Tek 1989). The density difference between CO_2 and CH_4 plays the most vital role in the separation of the gases. Fig. 5.4a and 4b show the density and viscosity of the CO_2-CH_4 mixture gases at the temperature of 40 °C, respectively. The properties were calculated by using the WebGasEOS v.2.01 developed by the Lawrence Berkeley National Laboratory (Reagan 2012). In comparison with CH_4, the higher density of CO_2 would lead to downward sinking. As depicted in Fig. 5.4a, the density of the mixed gases is closely related to the gas composition and pressure condition. Particularly, the density decreases significantly with addition of small amounts the CH_4 into CO_2. Such decreasing behavior slows down with increasing CH_4 fraction. It should be pointed out that the sudden increase of density occurs near the critical pressure of CO_2. This is result from that the CO_2 begins to become supercritical fluid at the condition.

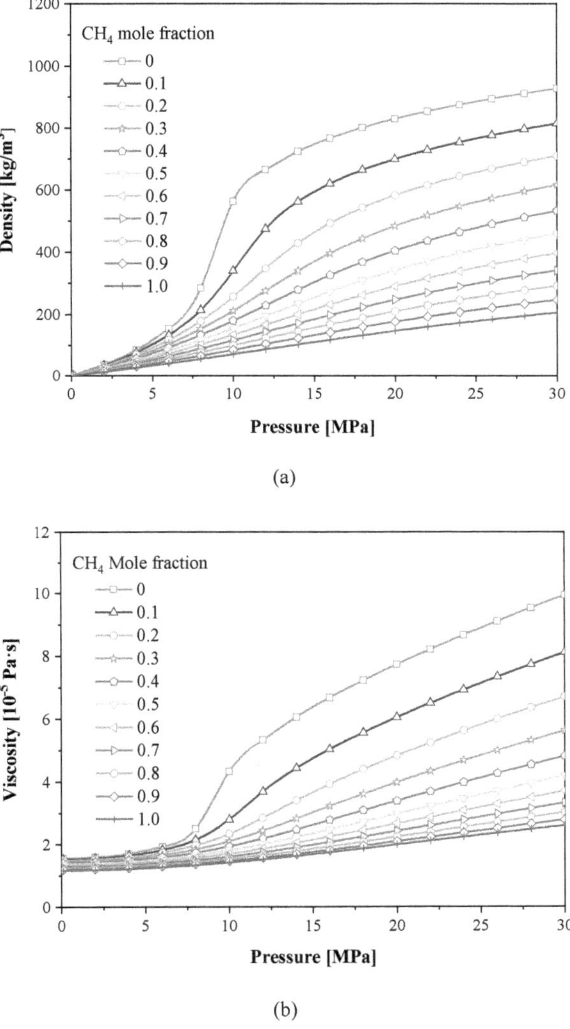

Figure 5.4 (a) Density of the CO_2-CH_4 mixtures at 40 °C; (b) viscosity of the CO_2-CH_4 mixtures at 40 °C (Cao et al. 2020)

However, the sudden change of the density of CO_2-CH_4 mixture gases disappears when the concentration of CH_4 is higher than approximately 30%. There are three potential reasons account for that. Firstly, increasing concentration of CH_4 would decreases the density of the mixture gases because of the low density of CH_4. Secondly, the CH_4 will take more partly pressure than CO_2 under compression because CH_4 is lowly comprehensible compared with CO_2. Thirdly, there is a Widom region for supercritical fluids, where some physicochemical properties including the density and compressibility exhibit anomalous behavior such as the sudden change (Imre et al. 2014; Velmovszki et al. 2019). The exits of contaminant such as CH_4 in this case would alter the location and shape of the Widom region (Raju et al. 2017), and affect the properties of the mixture gases.

Fig. 5.4b exhibits the dynamic viscosities of CO_2-CH_4 mixtures, whose tendency is similar as that of density in Fig. 5.4a. The mobility differences in CO_2-CH_4 displacement are primarily determined by the different dynamic viscosities. It is worth to mention that the difference of viscosities between CO_2 and CH_4 is beneficial for limiting the gas mixing (Oldenburg et al. 2001), but it may result in an unstable contact interface (Ma et al. 2019).

The dispersion coefficient of CO_2 in CH_4 normally ranges from 0.01 to 0.3 cm^2/min (Seo 2004), which is a relatively slow velocity compared with the advective and convective transport. It should be noted that the diffusion effect is proportional to the concentration gradient, thus its impact will decrease along with the mixing process of the gases. The mechanical dispersion is determined by the movement of the formation fluids (Tek 1989).

5.2.2 Spatial distribution of CO_2 and CH_4

Fig. 5.5 depicts the spatial distribution of CO_2 for the UGSR with 10% CO_2 as the cushion gas at the end of the CH_4 production stage, well shutting stage, CH_4 injection stage, and the second well shutting stage, i.e., the time of 120 d, 155 d, 335 d, and 365 d in the 1st, 5th, 10th, and 15th year, respectively. It shows that the CO_2 is concentrated at the bottom of the reservoir because of its larger density compared with CH_4. The concentration of CO_2 in the reservoir decreases gradually with time, due to the production of CO_2 accompanied with CH_4 and the mixing of CO_2 with CH_4. Specifically, the maximum concentration of CO_2 in the UGSR decreases from 100% to 92.3%, 68.7%, 53.5%, and 43.6% at the end of the 1st, 5th, 10th, and 15th year, respectively.

Fig. 5 also shows the distribution of CO_2 changes periodically over time corresponding to the operation scenarios (Fig. 5.3). In the first stage (CH_4 production), CO_2 moves towards the production well because of the pressure gradient. Meanwhile, the CO_2 and CH_4 in the reservoir are significantly mixed until the termination of gas production. In the second stage (well shutting), the mixing behavior of the gases is dominantly determined by the density difference and the diffusion effect. However, the influence on mixing is still minor because of the relatively low diffusion coefficient and short period of well shut-in. In the third stage (CH_4 injection), CO_2 migrates towards the CO_2 injection well under the displacement effect of CH_4 injection. The mixed zone of gases become smaller during this stage. The last stage, i.e., the second well shutting stage, also has a negligible effect on the mixing of the gases.

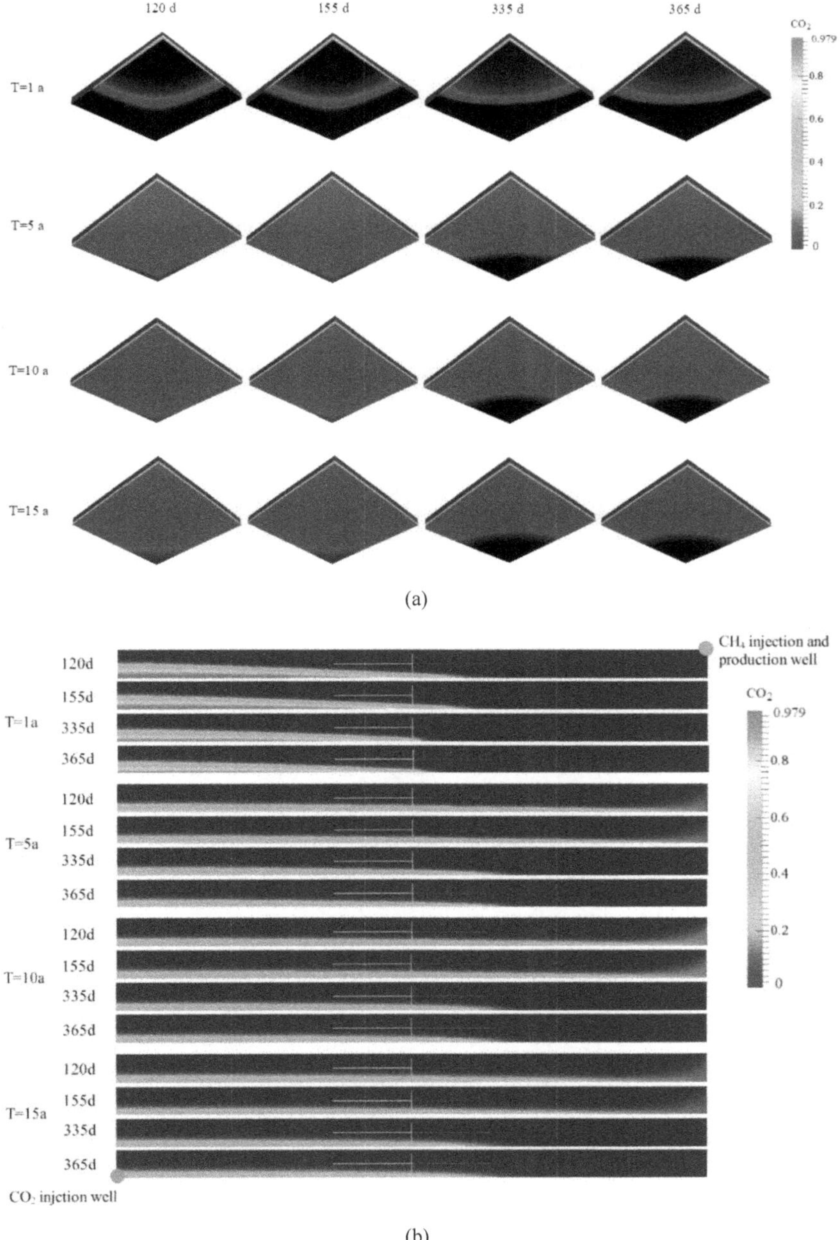

Figure 5.5 (a) CO_2 saturation in the 3D model; (b) CO_2 saturation in gas phase in the diagonal section which consists the CO_2 injection well and the CH_4 injection and production well (Cao et al. 2020)

5.3 Impacts of geological and engineering parameters

5.3.1 CO_2 fraction in the UGSR

To investigate the suitability of using CO_2 as cushion gas, the UGSR with a CO_2 concentration varying from 0 to 20% for the cushion is analyzed. The reservoir average pressure against time is summarized in Fig. 5.6. It shows that all the reservoir average pressure changes periodically over time, decreasing in the production stage, remaining constant during well shutting stage, and increasing in the injection stage. In the first cycle (Fig. 5.6b), the minimum value of reservoir pressure decreases with the increasing CO_2 concentration because of the higher compressibility of CO_2 compared with CH_4. With the operation of the UGSR, the maximum reservoir average pressure increases gradually as the produced CO_2 is replaced by the injected CH_4 that has a lower compressibility.

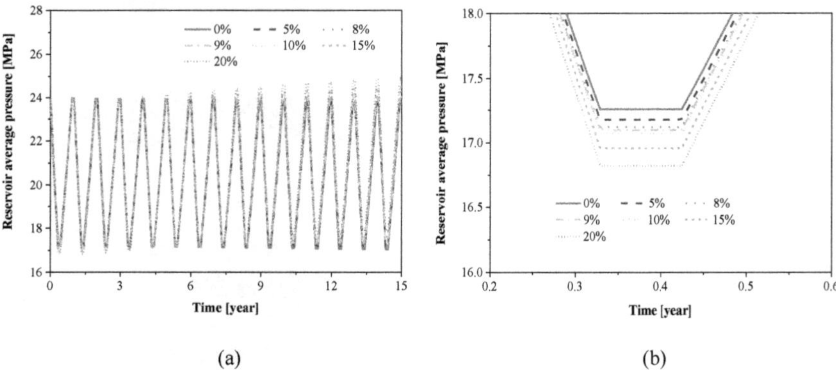

Figure 5.6 (a) Average reservoir pressure for the UGSR with different concentration of cushion gas CO_2; (b) the partially enlarged view (Cao et al. 2020)

To quantify the spatial distribution of the mixing region, the mixed thickness is used to characterize the interface between the working and cushion gases (Curtis et al. 2013). In this work, the mixed thickness was defined as the distances between the CO_2 fraction level of 10% and 90% along the diagonal of the reservoir that crossing both the CO_2 injection well and the CH_4 production well (see in Fig. 5.7). The mixed thickness for the UGSR with different concentration of CO_2 cushion are shown in Fig. 5.8. It is apparent that the thickness of the mixed zone increases with the increasing of the CO_2 initiated concentration, which can be explained by the following two reasons. Firstly, more volume of CO_2 means higher chance for the diffusion and mixing with CH_4. Secondly, the reservoir pressure of the UGSR with a higher concentration of CO_2 is relatively low (Fig. 5.6), which lead to a relatively higher diffusion coefficient (Liu et al. 2020).

The curve of the mixed zone's thickness for the UGSR with initial CO_2 concentration ranging from 8% to 10% is similar, in which the representative curve for the initial 9% of CO_2 can be divided into four stages over the lifetime of project (see in Fig. 5.9c). The distance between the CO_2 injection well and the point with a CO_2 concentration of 10% and 90% CO_2 along the same diagonal line are denoted as $r_{0.1}$ and $r_{0.9}$ and plotted in Fig. 5.9a and 5.9b against time, respectively. It can be seen that the change of $r_{0.1}$ is harmonized with the periodical operation scenarios (Fig. 5.3). This may potentially demonstrate that the $r_{0.1}$ is dominantly driven by the gas displacement effect. It should be pointed out that the value of $r_{0.1}$ is limited narrowly range from 189.9 to 193.7 m. Hence, it can be inferred that the abrupt change of mixed zone thickness ($r_{0.1}$- $r_{0.9}$) is mostly caused by the variation of $r_{0.9}$.

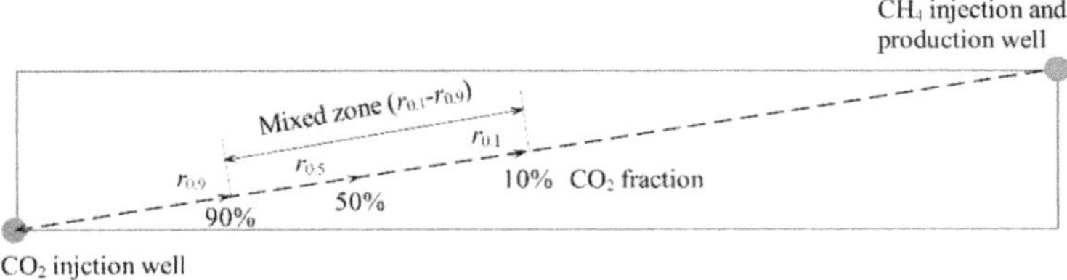

Figure 5.7 Schematic diagram of the mixed zone in the diagonal section of the reservoir (Cao et al. 2020)

The stage I corresponds to the CH_4 injection in the first cycle. In this stage, the $r_{0.1}$ increases gradually (Fig. 5.9a), while the $r_{0.9}$ decreases very fast from 123.5 to 64.4 m (Fig. 5.9b), thus the thickness of the mixed zone increases dramatically because of the mixing of CO_2 and CH_4 driven by the pressure gradient and diffusion effect. In stage II, the $r_{0.9}$ decreases only by 4.5 m and the $r_{0.1}$ slightly decreases by around 2 m, thus the mixed thickness changes only 2.5 m in 245 days during this stage, which corresponds to the period from the first well shutting to the second well shutting stage in the first operation cycle. Therefore, the mixed thickness in the second stage behaves smoothly. Although the change of the $r_{0.9}$ is not significant, the CO_2 concentration near the CO_2 injection well decreases gradually. With the continuously mixing of CO_2 and CH_4, the maximum CO_2 concentration in the UGSR decreases to a value lower than 90%, thereby the $r_{0.9}$ decreases to 0 and then the distance of the mixed zone substantially increases again within a short time, which is assigned to stage III. In stage IV, the distance of the mixed zone equals the $r_{0.1}$ and changes cyclically with the injection and production of CH_4.

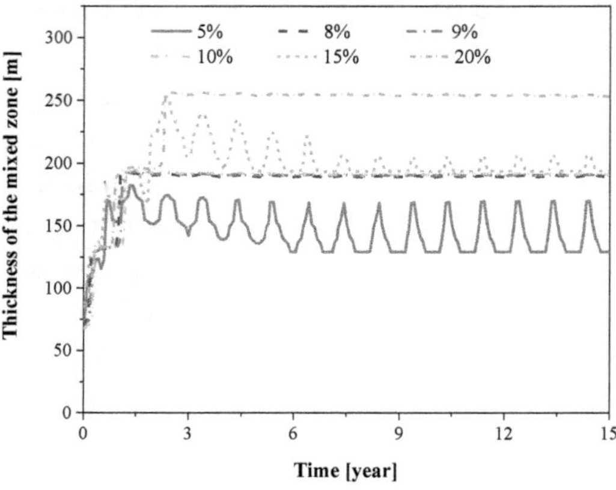

Figure 5.8 Thickness of the mixed zone for the UGSR with different concentration of CO_2 as cushion gas (Cao et al. 2020)

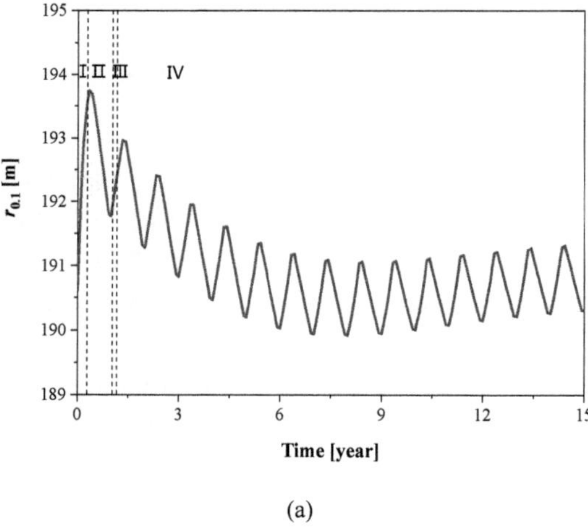

(a)

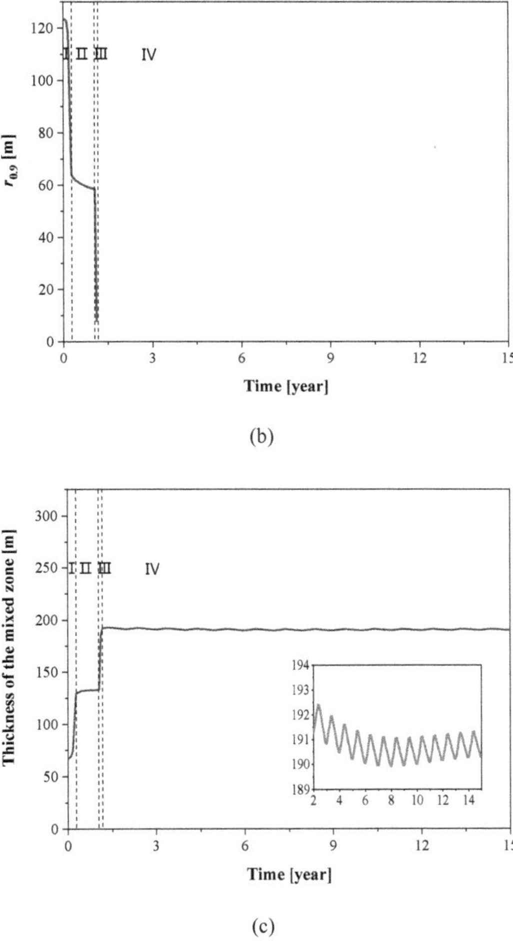

Figure 5.9 (a) The distance between the CO_2 injection well and the point with CO_2 concentration of 10% in the diagonal of the reservoir during production; (b) The distance between the CO_2 injection well and the point with CO_2 concentration of 90% in the diagonal line of the reservoir; (c) The thickness of the mixed zone during production (Cao et al. 2020)

Unlike the curve of $r_{0.9}$ for the UGSR with 10% CO_2 (Fig. 5.9b), there are two stable sections for the case with 20% CO_2 (Fig. 5.10), which corresponding to the two stable sections of the curve of the mixed zone (Fig. 5.8). Fig. 5.11 depicts the CO_2 concentration in the diagonal of the reservoir for the UGSR with 20% CO_2. It shows that the mixed zone thickness maintains at 131.1 m from 0.35 to 0.84 year, corresponding to the first stable section in Fig. 5.8. During this period, the CO_2 concentration in the region away from the CO_2 injection well gradually decreases to the threshold value of 90%, then resulting an expansion of the mixed zone. It also shows that the mixed zone thickness changes only from 189.7 to 195.9 m during the period of 0.95 to 2.22 years. This is corresponding to the second stable section in Fig. 5.8. During this period, the CO_2 concentration in the region close to CO_2 injection well decreases to lower than 90% gradually, then resulting an expansion of the mixed zone again.

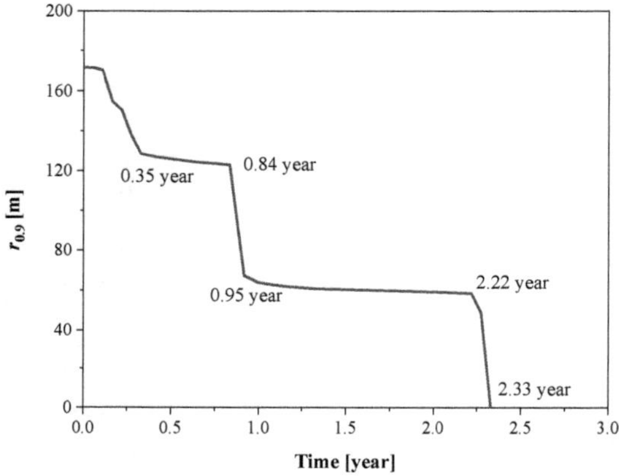

Figure 5.10 The relationship of $r_{0.9}$ with time for the UGSR with 20% CO_2 as cushion gas (Cao et al. 2020)

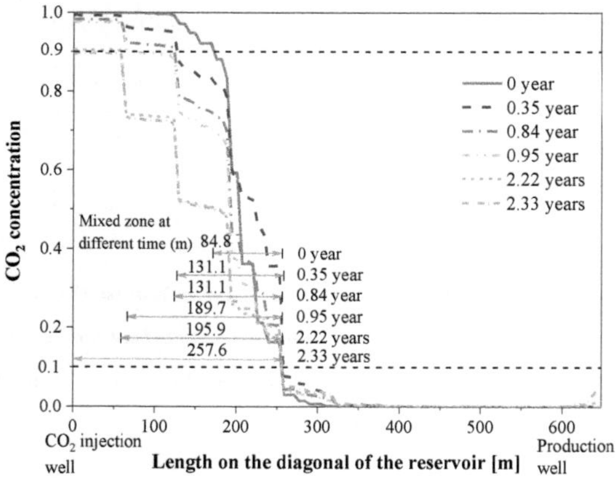

Figure 5.11 The CO_2 concentration in the diagonal of the reservoir for the UGSR with 20% CO_2 (Cao et al. 2020)

In every production cycle, the CO_2 concentration in the produced gas mixture (CO_2-CH_4) over 15 years is depicted Fig. 5.12. Clearly, the CO_2 concentration in the produced gases increases in every production cycle and reaches up to the maximum value at the termination of each production stage. The maximum CO_2 concentration increases sharply, when the CO_2 migrates to the production well in the first few years. Thereafter, the maximum concentration of CO_2 decreases gradually. This is result from the decreasing volume of CO_2 in the UGSR caused by production. The maximum value of the produced CO_2 concentration in the whole operation process is found at an earlier time with the increasing of the CO_2 cushion gas. Particularly, it occurs at the 6th and 10th year for the UGSR with CO_2 cushion gas of 20% and 9%, respectively. According to the National natural gas standards of China, the CO_2 concentration in the first class natural gas for civil fuel is not allowed higher than 3% (National standards of the P.R. China 2018). Therefore, it can be found that the optimal CO_2 concentration for the cushion gas in this operation scenario is 9%, which will be used as the base case in following studies.

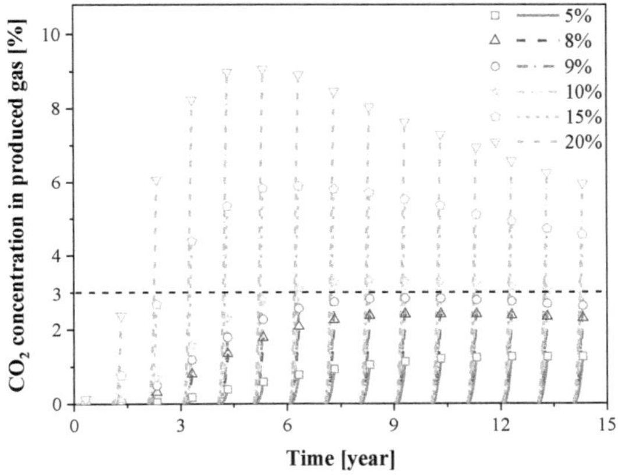

Figure 5.12 CO_2 concentration in the produced gas for the UGSR with different concentrations of CO_2 as cushion gas (Cao et al. 2020)

5.3.2 Reservoir temperature

Figs. 5.13-5.15 depict the average reservoir pressure, mixed zone thickness, and CO_2 concentration in produced gas over 15 years of operation under different temperatures, respectively. As expected, temperature has a positive impact on the reservoir pressure change, and the cyclical changes are synchronized with the operation scenarios (Fig. 5.3). Fig. 5.14 shows that the thickness of the mixed zone decreases with the increasing temperature in the initial stage, which is consistent with the results obtained by Ma et al. (2019). This is result from that the high reservoir pressure caused by high temperature increases the dynamic viscosity of the gases. Likewise, the rapidly increasing stage (stage III in Fig. 5.9) occurs earlier with a lower temperature. After that, the ranking of the mixed zone thickness reverses eventually in stage IV. During this stage, the mixing region increases with increasing temperature because of the high diffusion coefficient, which is demonstrated by the experiments conducted by Liu et al. (2020). Fig. 5.15 shows that the influence of temperature on the CO_2 concentration in the production well. It shows that the impact of temperature on the CO_2 concentration in the produced gas can be neglectable within the range from 30 to 50 °C.

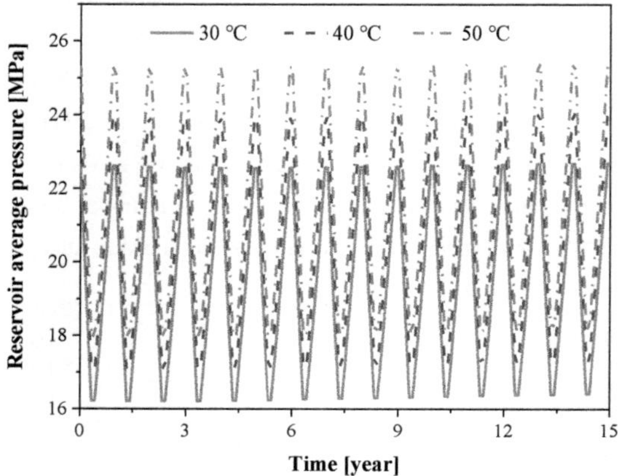

Figure 5.13 Reservoir average pressure for different temperature (Cao et al. 2020)

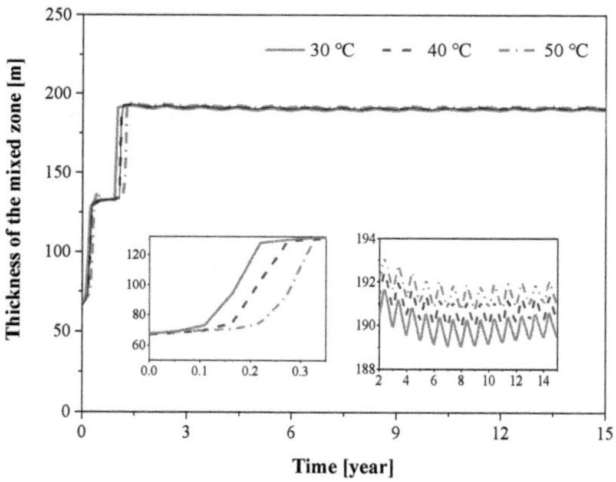

Figure 5.14 Thickness of the mixed zone for different temperature (Cao et al. 2020)

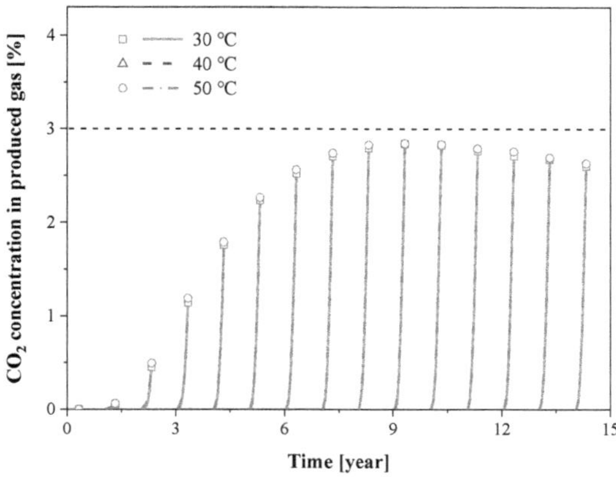

Figure 5.15 CO_2 concentration in the produced gas for different temperature (Cao et al. 2020)

5.3.3 Reservoir permeability

The permeability of depleted gas reservoirs may range from a few to hundreds of millidarcy. To quantify the effect of reservoir permeability on the mixing behavior of the gases, the UGSR with the reservoir horizontal permeability of 50, 70, 100, and 120 mD respectively was applied for simulation. As depicted in Fig. 5.16, the reservoir permeability has limited effect on the formation pressure, but the impact on the mixing behavior of CO_2 and CH_4 is much more significant. Fig. 5.17 reveals that the mixing thickness is positively related to the reservoir permeability while this relationship is inversed after entering the stage IV. During the CH_4 production stage in the first year (stage I in Fig. 5.9c), the reservoir with high permeability would accelerate the migration of CO_2, further promoting the mixing of CO_2 and CH_4. With continuous operation for the UGSR, more CO_2 would be produced along with CH_4 production in the scenario with high permeability, leading to the boundary with CO_2 concentration of 10% quickly approach the CO_2 injection well. In other words, the $r_{0.1}$ and the corresponding mixed zone thickness decrease quickly in the scenario with high permeability. Similarity to that mentioned above, the mixed zone thickness changes periodically with the production and injection of CH_4. The gases in the UGSR are much easier to be mixed in the reservoir with high permeability. In the stage IV (Fig. 5.9c), the mixing zone is dominantly determined by the $r_{0.1}$, the displacement effect of CH_4 injection is more pronounced in the permeable reservoir, thus the mixing zone thickness decreases with the increasing permeability and varies in a large amplitude.

The CO_2 concentrations in the produced gases for different reservoir horizontal permeability are compared in Fig. 5.18. As expected, CO_2 concentration in the produced gas is positively related to the reservoir permeability. It can be seen that the CO_2 concentration in the produced gases between 3 to 8 years for the UGSR with the permeability of 100 and 120 mD is very high, indicating high CO_2

production rates, which affects the distribution of CO_2 in the reservoir a lot and leads to the non-uniform fluctuation of the mixed region as depicted in Fig. 5.17. Therefore, the CO_2 concentration in a highly permeable reservoir may be lower than that of lowly permeable reservoir in the late operation period without re-injected CO_2 for supplementation. Likewise, the maximum CO_2 concentration in the produced gases is found at an earlier time in the highly permeable reservoir.

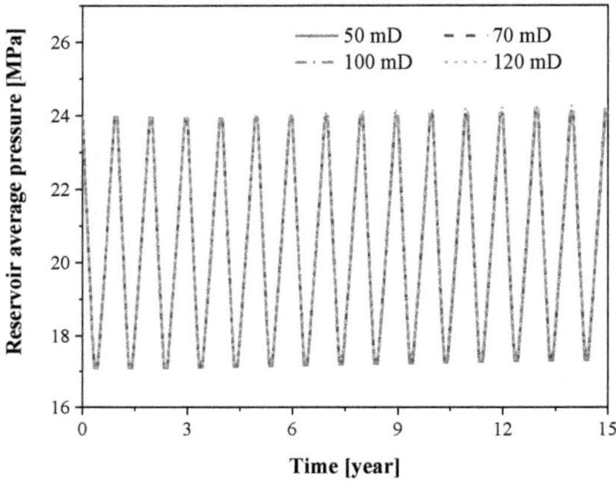

Figure 5.16 Reservoir average pressure for different reservoir horizontal permeability (Cao et al. 2020)

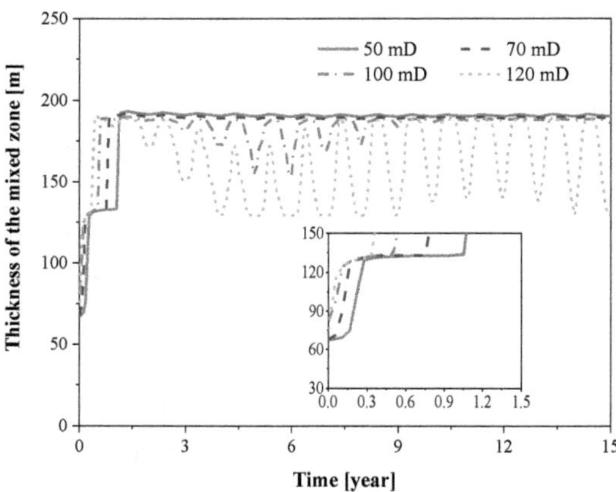

Figure 5.17 Thickness of the mixed zone for different for different reservoir horizontal permeability (Cao et al. 2020)

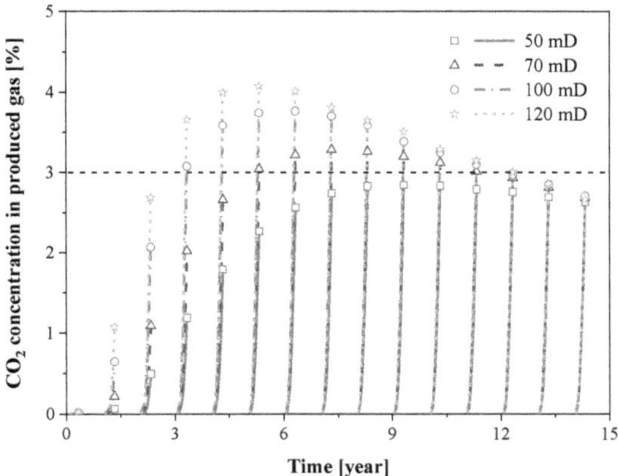

Figure 5.18 CO_2 concentration in the produced gas for different reservoir horizontal permeability (Cao et al. 2020)

5.3.4 Residual water saturation

Fig. 5.19 shows the reservoir average pressure for the UGSR with a residual water saturation of 15%, 20%, 25%, and 30%, respectively. Generally, the reservoir pressure changes periodically along with the CH_4 production and injection. It can be seen that the reservoir pressure increases with the rising residual water saturation, because the uncompressible water can build up high pressure in the reservoir. Fig. 5.20 depicts the thickness of the mixed zone. During the production stage in the first year (stage I in Fig. 5.9c), the mixing zone thickness in the reservoir tends to increase with the growing residual water saturation. This can be elaborated by the following reasons. The first one is that the occupied pores promote the injected CO_2 traveling deeper into the reservoir, resulting higher mixing thickness under the same injection volume. Another reason is that the residual water would generate more tortuous flow-paths for the gases as depicted in Fig. 5.21, which was discovered according to the result of nuclear magnetic resonance during a core flood experiment (Honari et al. 2016). Therefore, the rapidly increasing stage (stage III in Fig. 5.9c) occurs earlier for the scenario with a higher residual water saturation.

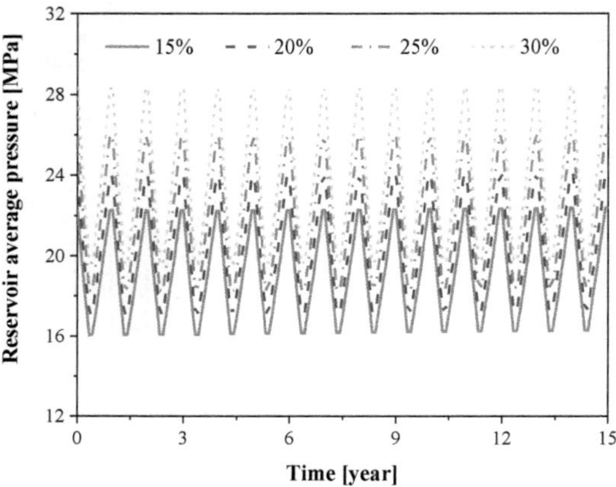

Figure 5.19 Reservoir average pressure for different residual water saturation (Cao et al. 2020)

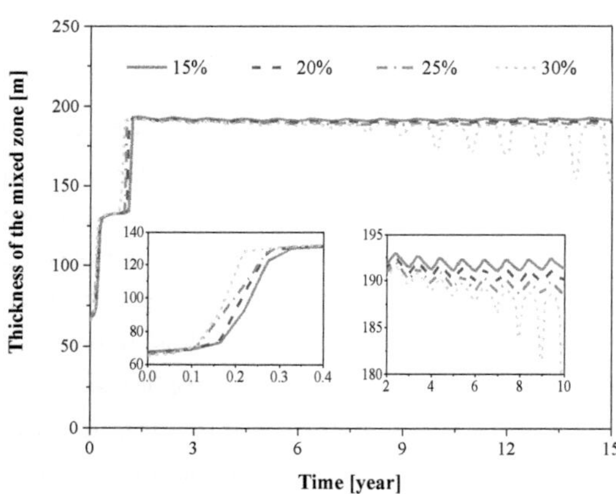

Figure 5.20 Thickness of the mixed zone for different residual water saturation (Cao et al. 2020)

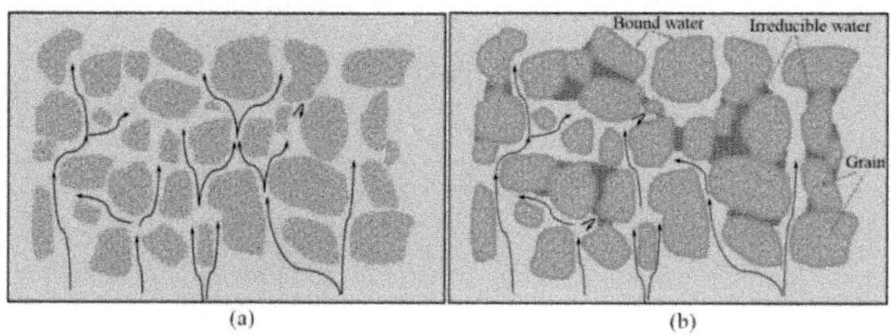

Figure 5.21 (a) Fluid flow in unsaturated porous rock; (b) Fluid flow in water saturated porous rock (Honari et al. 2016)

Similarity to the influence of the permeability on the mixed zone thickness, it is inversely correlated to the residual water saturation in stage IV. This is result from the dissolution of CO_2 in residual water. As presented in Fig. 5.22, less CO_2 exists as gaseous in the UGSR with higher water saturation, because more CO_2 dissolved in the residual water. Fig. 5.22 also depicts the mass of dissolved CO_2 changes periodicity in a certain range during the operation of the UGSR, because the dissolution of CO_2 is principally dependent on the pressure and temperature. The CO_2 concentration in the produced gas for different residual water saturation is illustrated in Fig. 5.23. It shows that the impact of the residual water on the CO_2 concentration in the produced gas is still minor, which is result from that the difference of the gaseous CO_2 in the UGSR is not very significant as depicted in Fig. 5.22.

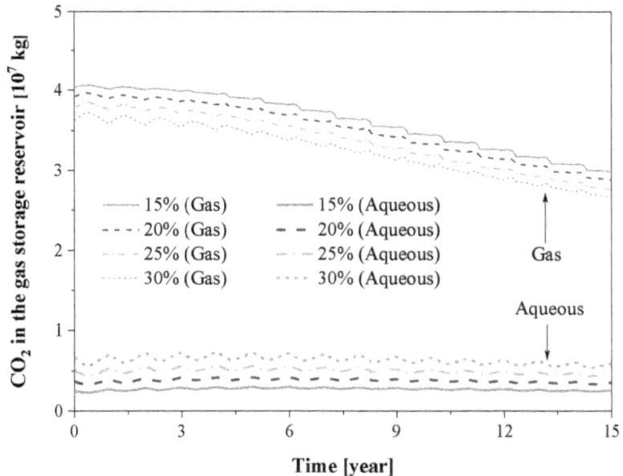

Figure 5.22 CO_2 in gaseous and aqueous in the UGSR for different residual water saturation (Cao et al. 2020)

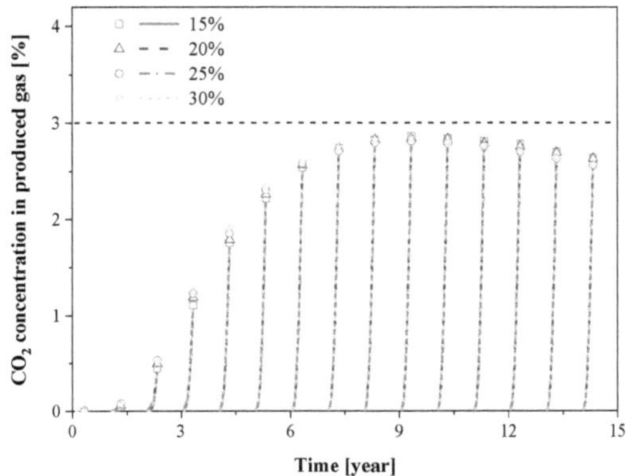

Figure 5.23 CO_2 concentration in the produced gas for different residual water saturation (Cao et al. 2020)

5.3.5 CH_4 production rate

The average reservoir pressure of the UGSR with different production rates are plotted against time in Fig. 5.24. Clearly, the average reservoir pressure changes periodically and decreases with the increasing production rate. The effect of the CH_4 production rate on the mixed zone thickness is shown in Fig. 5.25. The impact of the production rate on the mixing behavior of CO_2 and CH_4 is limited, particularly in the early operation stages I-III. Considering that in the case of determined CH_4 production rate (determined pressure gradient), the effect of temperature and residual water saturation that affect the diffusion effect on the mixed zone thickness is significant (Fig. 5.14 and 5.20), it can be inferred that the mixing of CO_2 and CH_4 during these stages is dominantly controlled by the diffusion effect rather than pressure gradient. In the subsequent stage IV, the mixing zone thickness has a slight decrease with the increasing production rate because of the impact of the pressure gradient. As depicted in Fig. 5.26, the production rate has a positive correlation with the CO_2 concentration in the produced gases. Fig. 5.26 also shows that the maximum concentration of produced CO_2 is limited to under 3%, which still satisfies the required Standard (National standards of the P.R. China 2018). Therefore, the CH_4 production rate of 4.2 kg/s could be used in this case, to improve the efficiency of the UGSR.

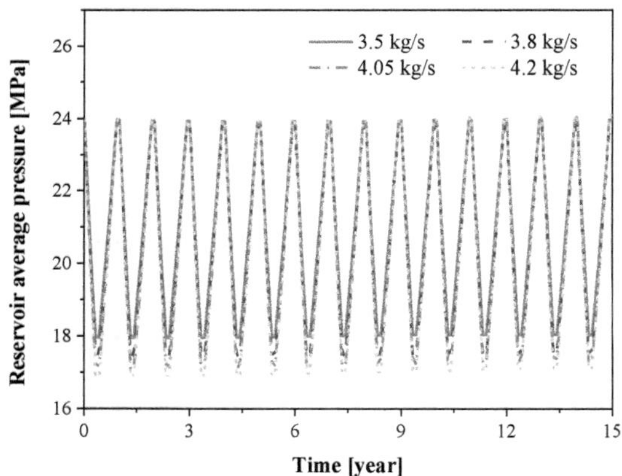

Figure 5.24 Reservoir average pressure for different production rate (Cao et al. 2020)

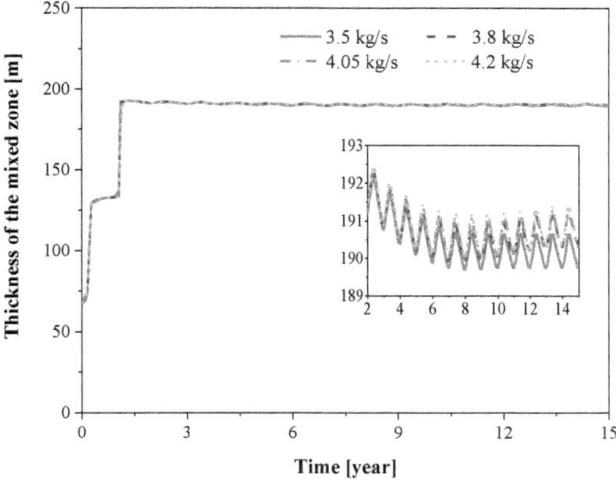

Figure 5.25 Thickness of the mixed zone for different for different production rate (Cao et al. 2020)

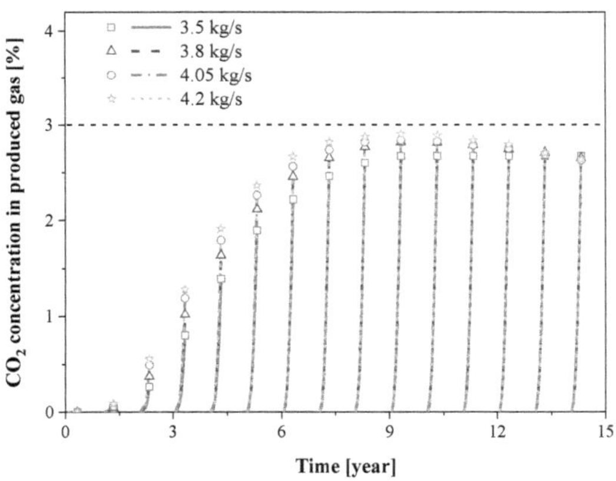

Figure 5.26 CO_2 concentration in the produced gas for different production rate (Cao et al. 2020)

5.4 Summary

The Donghae gas reservoir was used to investigate the suitability of using CO_2 as cushion gas for UGSR in depleted gas reservoir. The maximum concentration of CO_2 can be used for cushion gas is 9% under the condition of production and injection for 120 and 180 days in a production cycle at a rate of 4.05 and 2.7 kg/s, respectively.

The typical curve of the mixing zone thickness could be divided into four stages, including the increasing stage, the smooth stage, the suddenly increasing stage, and the periodic change stage. It should be pointed out that there exists two smooth stage and suddenly increasing stages when the CO_2

concentration in the cushion gas increases to 20%. During the periodic change stage, the mixed zone thickness increases with the growing of the CO_2 fraction, temperature, production rate, and the decreasing of permeability and water saturation. The correlation of the mixing zone thickness with the reservoir temperature, permeability, and residual water was inverse in the former stages.

The CO_2 concentration in the UGSR, reservoir permeability, and production rate have significant impact on the breakthrough of CO_2 in the production well, while the influence of the water saturation and temperature are neglectable. Generally, CO_2 can be used as cushion gas in the UGSR to improve the cost-effectiveness of CCS. For the purpose of utilizing more CO_2 as cushion gas in the UGSR, CO_2 should be injected for supplementation during the operation of UGSR.

6 Conclusions and outlook

The atmospheric CO_2 emissions is the main contributor to the continuous rise in global warming and climate change. To deal with such intense global climate problem, carbon capture and storage (CCS) is considered as the most promising technology for slowing down the atmospheric CO_2 emissions. However, the contribution of CCS is still very limited until now in mitigating climate change because of the related risks and the lack of economic incentives, which challenge the large-scale application of CCS. In this thesis, regarding the risks associated with CCS, a parametric uncertainty analysis for CO_2 storage was conducted and the general role of different key geomechanical and hydrogeological parameters in response to CO_2 injection was determined. To increase the cost-effectiveness of CCS operation, co-injection of CO_2 with impurities associated with enhanced gas recovery (CSEGR) and the utilization of CO_2 as cushion gas in the underground gas storage reservoir (UGSR) were proposed and analyzed.

To characterize the complex THM process of CO_2 storage, the fundamentals of the coupled THM processes were introduced in chapter 2. The mass and energy transport during the multiphase and multicomponent flow were presented. The governing equations of the mechanical behavior of formation rocks such as the constitutive equations were also introduced in detail. Moreover, the coupled THM simulator TOUGH2MP (TMVOC)-FLAC3D was developed based on the multiphase and multicomponent flow code TOUGH2MP and the geomechanical software FLAC3D.

By using the developed TOUGH2MP (TMVOC)-FLAC3D simulator, the parametric uncertainty analysis for CO_2 storage was conducted in chapter 3. A generic 2D model with geometric parameters that are averaged from worldwide large-scale CCS projects is developed and used for simulation. The key geomechanical and hydrogeological parameters, including the Young's modulus, Poisson's ratio, Biot coefficient, permeability, anisotropy ratio, and the injection rate were compiled from the literature and sampled randomly based on the Quasi-Monte Carlo method, to quantitatively analyze the role of uncertainty parameters within a possible range during CO_2 injection. Numerical simulation for the sampled data was performed using the developed TOUGH2MP (TMVOC)-FLAC3D simulator. Based on the simulation results, the general role of different geomechanical and hydrogeological parameters on the formation response due to CO_2 injection was determined by using the distance correlation. The results showed that the reservoir Young's modulus plays the most crucial role in formation deformation including vertical displacement. Furthermore, the reservoir permeability and the injection rate are the two most vital parameters in determining the pressure change. A risk factor with the physical properties of the caprock and the corresponding stress conditions considered was defined to characterize the risk of the caprock failure associated with CO_2 injection in CCS operation. The analysis showed that the pressure change exhibits a much closer correlation with the risk factor in comparison to the formation deformation, demonstrating the significance of pressure change in the integrity assessment of the caprock. In addition, based on the machine learning approach in support vector regression (SVR), the SVR surrogate model was well-trained on the basis of the dataset gained from the simulated results. The reliability of the SVR surrogate model was verified through the test dataset. Thereafter, the formation response to CO_2 injection, including the formation deformation as well as pressure change, can be predicted within a very short time by using the trained SVR surrogate model. The rules obtained from the predicted results and the distance correlation analysis were confirmed and verified based on a comparison to each other. The methods and working scheme used in this work can be applied to guide time and effort spent in reducing the uncertainty in these parameters, which is beneficial for the acquiring of trustworthy model forecasts and risk assessments in CCS projects.

For the purpose of cutting the cost of CCS operation, co-injection of CO_2 with impurity gas into the depleted gas reservoir for storage associated with the potential enhanced gas recovery was investigated in chapter 4. The main impurities from the flue gas that is the main source of CO_2 emissions, i.e., N_2 and O_2, were considered with three different concentrations. Both the scenarios of dedicated for CO_2 storage and CSEGR were taken into consideration. In the scenario of dedicated for CO_2 storage, the gas was injected into the depleted gas reservoir until the average reservoir pressure recover to the original reservoir pressure. In the scenario of CSEGR, CO_2-EGR was conducted firstly under the condition of maintaining the well bottom pressure constant. The operation of CO_2-EGR was terminated when the concentration of impurities in the produced gas reach to a threshold value, then the unit was transformed for CO_2 storage. The impacts of the residual CH_4 content, reservoir temperature, residual water saturation, and injection rate on the performance of CO_2 storage and CSEGR were analyzed in detail. The results showed that the CO_2 contained the impurities of N_2 and O_2 is feasible to be injected into depleted gas reservoirs for CO_2 storage and CSEGR. The effect of the impurities on the CO_2 storage capacity is dependent on the concentration of the impurities and the reservoir pressure and temperature conditions. The depleted gas reservoir with a relatively low temperature and low irreducible water saturation is favorable for the CO_2 storage capacity. The depleted gas reservoir with a low primary gas recovery is suitable for CSEGR, while it is favorable for dedicated CO_2 storage when the primary gas recovery is high. The N_2 and O_2 may migrate faster than CO_2 in the depleted gas reservoir and concentrate at an area away from the injection well, which is so called chromatographic partitioning phenomenon. This phenomenon may occur in both gaseous and aqueous phases when N_2 and O_2 were co-injected with CO_2 into the depleted gas reservoirs. The chromatographic partitioning phenomenon can be applied as a monitoring method for the CO_2 front and potential CO_2 leakage. Specifically, the detection of N_2 and O_2 front can be considered as a signal of CO_2 front and potential CO_2 leakage after a time lag. In addition to the solubility and concentration of the impurity gas affecting this phenomenon, there is a critical water saturation for the occurrence of significant chromatographic partitioning phenomenon. Generally, the N_2 and O_2 can be co-injected with CO_2 into depleted gas reservoirs to decrease the cost on gas separation and get additional economic profit through CSEGR.

In an attempt to increase the cost-effectiveness of CCS, the suitability of utilizing CO_2 as cushion gas in the UGSR was analyzed in chapter 5. The geological parameters of Donghae depleted gas reservoir that is located in Ulleung basin in Korea was used as an example. After the injection of CO_2 as cushion gas, the cyclic CH_4 production and injection for 120 and 180 days at a rate of 4.05 and 2.7 kg/s respectively were conducted over a period of 15 years to acquire the mixing behavior of CO_2 and CH_4 in a relatively long-term period. Additionally, the impacts of the CO_2 fraction, reservoir temperature, reservoir permeability, residual water, and production rate on the mixing behavior of the CO_2-CH_4 system and gas production were examined in detail. The results showed that the maximum CO_2 concentration that can be used for cushion gas is 9% under the geological and operation conditions. The typical curve of the mixing zone thickness can be divided into four stages, including the increasing stage, smooth stage, suddenly increasing stage, and periodic change stage. In the periodic change stage, the mixed zone thickness has a positive correlation with the CO_2 fraction, temperature, production rate, while it has a negative correlation with the reservoir permeability and water saturation. It should be mentioned that the correlation of the mixing zone thickness with reservoir temperature, permeability, and residual water was inverse in the former stages. The CO_2 fraction in the UGSR, reservoir permeability, and production rate affect the breakthrough of CO_2 in the production well significantly, while the impact of water saturation and temperature is neglectable. In general, CO_2 can be used as cushion gas in the UGSR to improve the cost-effectiveness of CCS.

Overall, the parametric uncertainty analysis performed in this thesis is favorable for the risk assessments in CCS projects. Additionally, co-injection of CO_2 with impurities into depleted gas reservoirs associated with CSEGR and the utilization of CO_2 as cushion gas in UGSR are beneficial for improving the economic incentives of CCS operation. Therefore, this thesis is advantageous for promoting the application of CCS on large scale and mitigating the atmospheric CO_2 emissions.

In future work, to reduce the uncertainty and risks of CCS, the chemical reactions between injected CO_2 with the well equipment and formation rock are supposed to be taken into consideration. Especially, the long-term experiments and molecular dynamic simulations are needed to figure out the kinetics between CO_2 with the well string, cement, as well as formation minerals under the relevant conditions. Attempting to make the CCS assessment more intelligential, the machine learning technology is encouraged to be used. In addition, the overall economic analysis related to co-injection of CO_2 with impurities should be analyzed systemically. Last but not the least, it is suggested to conduct field experiments on co-injecting CO_2 with impurity gas for EGR and utilizing CO_2 as cushion gas in the UGSR.

References

Abba, M., Abbas, A., Saidu, B., Nasr, G., Al-Otaibi, A. (2018). Effects of gravity on flow behaviour of supercritical CO_2 during enhanced gas recovery (EGR) by CO_2 injection and sequestration. In *the Fifth CO2 Geological Storage Workshop*, Utrecht, The Netherlands, November 21-23.

Abba, M.K., Abbas, A.J., Nasr, G.G. (2017). Enhanced gas recovery by CO_2 injection and sequestration: Effect of connate water salinity on displacement efficiency. In *Abu Dhabi International Petroleum Exhibition & Conference*, Abu Dhabi, UAE, November 13-16.

Abdoulghafour, H., Gouze, P., Luquot, L., Leprovost, R. (2016). Characterization and modeling of the alteration of fractured class-G Portland cement during flow of CO_2-rich brine. *International Journal of Greenhouse Gas Control, 48*, 155-170.

Abid, K., Gholami, R., Choate, P., Nagaratnam, B.H. (2015). A review on cement degradation under CO_2-rich environment of sequestration projects. *Journal Of Natural Gas Science And Engineering, 27*, 1149-1157.

Abidoye, L.K., Khudaida, K.J., Das, D.B. (2015). Geological carbon sequestration in the context of two-phase flow in porous media: A review. *Critical Reviews in Environmental Science and Technology, 45*(11), 1105-1147.

Adams, E.E., Caldeira, K. (2008). Ocean storage of CO_2. *Elements, 4*(5), 319-324.

Al-Hasami, A., Ren, S., Tohidi, B. (2005). CO_2 injection for enhanced gas recovery and geo-storage: reservoir simulation and economics. In *SPE Europec/EAGE Annual Conference*, Maddrid, Spain, June 13-16.

Allen, R., Nilsen, H.M., Lie, K.A., Møyner, O., Andersen, O. (2018). Using simplified methods to explore the impact of parameter uncertainty on CO_2 storage estimates with application to the Norwegian Continental Shelf. *International Journal of Greenhouse Gas Control, 75*, 198-213.

Alpermann, T., Dietrich, M., Ostertag-Henning, C. (2016). Mineral trapping of a CO_2/H_2S mixture by hematite under initially dry hydrothermal conditions. *International Journal of Greenhouse Gas Control, 51*, 346-356.

Aminu, M.D., Nabavi, S.A., Rochelle, C.A., Manovic, V. (2017). A review of developments in carbon dioxide storage. *Applied Energy, 208*, 1389-1419.

Ampomah, W., Balch, R.S., Cathar, M., Will, R., Lee, S.Y., Dai, Z. (2016). Performance of CO_2-EOR and storage processes under uncertainty. In *SPE Europec featured at 78th EAGE Conference and Exhibition*, Vienna, Austria, May 30 - June 2.

Anchliya, A., Ehlig-Economides, C.A., Jafarpour, B. (2012). Aquifer management to accelerate CO_2 dissolution and trapping. *Spe Journal, 17*(03), 805-816.

Arts, R.J., Chadwick, A., Eiken, O., Thibeau, S., Nooner, S. (2008). Ten years' experience of monitoring CO_2 injection in the Utsira Sand at Sleipner, offshore Norway. *First Break, 26*(1), 65-72.

Arts, R.J., Vandeweijer, V.P., Hofstee, C., Pluymaekers, M.P.D., Loeve, D., Kopp, A., Plug, W.J. (2012). The feasibility of CO_2 storage in the depleted P18-4 gas field offshore the Netherlands (the ROAD project). *International Journal of Greenhouse Gas Control, 11*, S10-S20.

Aryana, S. A., Barclay, C., Liu, S. (2014). *North cross devonian unit - a mature continuous CO_2 flood beyond 200% HCPV injection.* In *SPE Annual Technical Conference and Exhibition*, Amsterdam, The Netherlands, October 27-29.

Atia, A., Mohammedi, K. (2018). A review on the application of enhanced oil/gas recovery through CO_2 sequestration. *Carbon Dioxide Chemistry, Capture and Oil Recovery*.

Audigane, P., Gaus, I., Czernichowski-Lauriol, I., Pruess, K., Xu, T. (2007). Two-dimensional reactive transport modeling of CO_2 injection in a saline aquifer at the Sleipner site, North Sea. *American journal of science, 307*(7), 974-1008.

Audigane, P., Gaus, I., Pruess, K., Xu, T. (2006). *A long term 2D vertical modelling study of CO_2 storage at Sleipner (North Sea) using TOUGHREACT*. In *Proceedings of TOUGH Symposium 2006*, Berkeley, California, May 15-17.

Azzolina, N.A., Peck, W.D., Hamling, J.A., Gorecki, C.D., Ayash, S.C., Doll, T.E., et al. (2016). How green is my oil? A detailed look at greenhouse gas accounting for CO_2-enhanced oil recovery (CO_2-EOR) sites. *International Journal of Greenhouse Gas Control, 51*, 369-379.

Bachu, S. (2007). Carbon dioxide storage capacity in uneconomic coal beds in Alberta, Canada: Methodology, potential and site identification. *International Journal of Greenhouse Gas Control, 1*(3), 374-385.

Bachu, S. (2013). Drainage and imbibition CO_2/brine relative permeability curves at in situ conditions for sandstone formations in western Canada. *Energy Procedia, 37*, 4428-4436.

Bachu, S. (2015). Review of CO_2 storage efficiency in deep saline aquifers. *International Journal of Greenhouse Gas Control, 40*, 188-202.

Bachu, S., Bennion, D.B. (2009). Chromatographic partitioning of impurities contained in a CO_2 stream injected into a deep saline aquifer: Part 1. Effects of gas composition and in situ conditions. *International Journal of Greenhouse Gas Control, 3*(4), 458-467.

Bachu, S., Hawkes, C., Lawton, D., Pooladi-Darvish, M., Perkins, E. (2009). CCS site characterisation criteria. *IEAG Greenhouse Gas R&D Programme (IEAGHG)*, 10.

Bachu, S., Pooladi-Darvish, M., Hong, H. (2009). Chromatographic partitioning of impurities (H_2S) contained in a CO_2 stream injected into a deep saline aquifer: Part 2. Effects of flow conditions. *International Journal of Greenhouse Gas Control, 3*(4), 468-473.

Bai, M., Zhang, Z., Fu, X. (2016). A review on well integrity issues for CO_2 geological storage and enhanced gas recovery. *Renewable and Sustainable Energy Reviews, 59*, 920-926.

Bao, J., Chu, Y., Xu, Z., Tartakovsky, A.M., Fang, Y. (2014). Uncertainty quantification for the impact of injection rate fluctuation on the geomechanical response of geological carbon sequestration. *International Journal of Greenhouse Gas Control, 20*, 160-167.

Bao, J., Hou, Z., Fang, Y., Ren, H., Lin, G. (2013). Uncertainty quantification for evaluating impacts of caprock and reservoir properties on pressure buildup and ground surface displacement during geological CO_2 sequestration. *Greenhouse Gases: Science and Technology, 3*(5), 338-358.

Baran, P., Zarębska, K., Krzystolik, P., Hadro, J., Nunn, A. (2014). CO_2-ECBM and CO_2 sequestration in Polish coal seam – experimental study. *Journal of Sustainable Mining, 13*(2), 22-29.

Barrufet, M.A., Bacquet, A., Falcone, G. (2010). Analysis of the storage capacity for CO_2 sequestration of a depleted gas condensate reservoir and a saline aquifer. *Journal of Canadian Petroleum Technology, 49*(08), 23-31.

Benisch, K., Graupner, B., Bauer, S. (2013). The coupled OpenGeoSys-eclipse simulator for simulation of CO_2 storage – code comparison for fluid flow and geomechanical processes. *Energy Procedia, 37*, 3663-3671.

Biagi, J., Agarwal, R., Zhang, Z. (2016). Simulation and optimization of enhanced gas recovery utilizing CO_2. *Energy, 94*, 78-86.

Bian, X.Q., Han, B., Du, Z.M., Jaubert, J.N., Li, M.J. (2016). Integrating support vector regression with genetic algorithm for CO_2-oil minimum miscibility pressure (MMP) in pure and impure CO_2 streams. *Fuel, 182*, 550-557.

Bian, X.Q., Xiong, W., Kasthuriarachchi, D.T.K., Liu, Y.B. (2019). Phase equilibrium modeling for carbon dioxide solubility in aqueous sodium chloride solutions using an association equation of state. *Industrial & Engineering Chemistry Research, 58*(24), 10570-10578.

Birkholzer, J., Zhou, Q., Tsang, C. (2009). Large-scale impact of CO_2 storage in deep saline aquifers: A sensitivity study on pressure response in stratified systems. *International Journal of Greenhouse Gas Control, 3*(2), 181-194.

Bjørnarå, T.I., Bohloli, B., Park, J. (2018). Field-data analysis and hydromechanical modeling of CO_2 storage at In Salah, Algeria. *International Journal of Greenhouse Gas Control, 79*, 61-72.

Böhm, G., Carcione, J.M., Gei, D., Picotti, S., Michelini, A. (2015). Cross-well seismic and electromagnetic tomography for CO_2 detection and monitoring in a saline aquifer. *Journal of Petroleum Science and Engineering, 133*, 245-257.

Boot-Handford, M.E., Abanades, J.C., Anthony, E.J., Blunt, M.J., Brandani, S., Mac Dowell, N., et al. (2014). Carbon capture and storage update. *Energy Environ. Sci., 7*(1), 130-189.

Boreham, C., Underschultz, J., Stalker, L., Kirste, D., Freifeld, B., Jenkins, C., Ennis-King J. (2011). Monitoring of CO_2 storage in a depleted natural gas reservoir: Gas geochemistry from the CO2CRC Otway Project, Australia. *International Journal of Greenhouse Gas Control, 5*(4), 1039-1054.

Böser, W., Belfroid, S. (2013). Flow assurance study. *Energy Procedia, 37*, 3018-3030.

Bourne, S., Crouch, S., Smith, M. (2014). A risk-based framework for measurement, monitoring and verification of the Quest CCS Project, Alberta, Canada. *International Journal of Greenhouse Gas Control, 26*, 109-126.

Bradshaw, J., Bachu, S., Bonijoly, D., Burruss, R., Holloway, S., Christensen, N.P., Mathiassen, O.M. (2007). CO_2 storage capacity estimation: issues and development of standards. *International Journal of Greenhouse Gas Control, 1*(1), 62-68.

Bratley, P., Fox, B.L. (2003). Implementing Sobols quasirandom sequence generator (Algorithm 659). *ACM Transactions on Mathematical Software, 29*(1), 49-57.

Brewer, P.G., Friederich, G., Peltzer, E.T., Orr, F.M. (1999). Direct experiments on the ocean disposal of fossil fuel CO_2. *Science, 284*(5416), 943-945.

Brinckerhoff, P. (2011). Accelerating the uptake of CCS: industrial use of captured carbon dioxide. Retrieved from https://www.globalccsinstitute.com/resources/publications-reports-research/?search=Accelerating+the+Uptake+of+CCS, accessed on June 11, 2019.

Brown, D.W. (2000). A hot dry rock geothermal energy concept utilizing supercritical CO_2 instead of water. In *Proceedings of the twenty-fifth workshop on geothermal reservoir engineering*, Stanford University, Stanford, California, January 24-26.

Brush, R.M., Davitt, H.J., Aimar, O.B., Arguello, J., Whiteside, J.M. (2000). Immiscible CO_2 flooding for increased oil recovery and reduced emissions. In *SPE/DOE Improved Oil Recovery Symposium*, Tulsa, Oklahoma, USA, April 3-5.

Bui, M., Adjiman, C.S., Bardow, A., Anthony, E.J., Boston, A., Brown, S., et al. (2018). Carbon capture and storage (CCS): the way forward. *Energy & Environmental Science, 11*(5), 1062-1176.

Burges, C.J. (1998). A tutorial on support vector machines for pattern recognition. *Data Mining and Knowledge Discovery, 2*(2), 121-167.

Burnside, N.M., Naylor, M. (2014). Review and implications of relative permeability of CO_2/brine systems and residual trapping of CO_2. *International Journal of Greenhouse Gas Control, 23*, 1-11.

Burton, E., Beyer, J., Bourcier, W., Mateer, N., Reed, J. (2013). Carbon utilization to meet California's climate change goals. *Energy Procedia, 37*, 6979-6986.

Busch, A., Alles, S., Gensterblum, Y., Prinz, D., Dewhurst, D., Raven, M., et al. (2008). Carbon dioxide storage potential of shales. *International Journal of Greenhouse Gas Control, 2*(3), 297-308.

Cameron, D.A., Durlofsky, L.J. (2012). Optimization of well placement, CO_2 injection rates, and brine cycling for geological carbon sequestration. *International Journal of Greenhouse Gas Control, 10*, 100-112.

Cao, C., Liao, J., Hou, Z., Wang, G., Feng, W., Fang, Y. (2020). Parametric uncertainty analysis for CO_2 sequestration based on distance correlation and support vector regression. *Journal of Natural Gas Science and Engineering, 77*.

Cao, C., Liao, J., Hou, Z., et al. (2020). Numerical modeling for CO_2 storage with impurities associated with enhanced gas recovery in depleted gas reservoirs. Manuscript.

Cao, C., Liao, J., Hou, Z., Xu, H., Mehmood, F., Wu, X. (2020). Utilization of CO_2 as cushion gas for depleted gas reservoir transformed gas storage reservoir. *Energies, 13*(3).

Cao, C., Liu, H., Hou, Z., Mehmood, F., Liao, J., Feng, W. (2020). A review of CO_2 storage in view of safety and cost-effectiveness. *Energies, 13*(3).

Carcione, J.M., Gei, D., Picotti, S., Michelini, A. (2012). Cross-hole electromagnetic and seismic modeling for CO_2 detection and monitoring in a saline aquifer. *Journal of Petroleum Science and Engineering, 100*, 162-172.

Carpenter, S.M., Koperna, G. (2014). Development of the first internationally accepted standard for geologic storage of carbon dioxide utilizing enhanced oil recovery (EOR) under the International Standards Organization (ISO) Technical Committee TC-265. *Energy Procedia, 63*, 6717-6729.

Carroll, A.G., Przeslawski, R., Radke, L.C., Black, J.R., Picard, K., Moreau, J.W., et al. (2014). Environmental considerations for subseabed geological storage of CO_2: A review. *Continental Shelf Research, 83*, 116-128.

Celia, M.A., Bachu, S., Nordbotten, J.M., Bandilla, K.W. (2015). Status of CO_2 storage in deep saline aquifers with emphasis on modeling approaches and practical simulations. *Water Resources Research, 51*(9), 6846-6892.

Celia, M.A., Nordbotten, J.M., Court, B., Dobossy, M., Bachu, S. (2011). Field-scale application of a semi-analytical model for estimation of CO_2 and brine leakage along old wells. *International Journal of Greenhouse Gas Control, 5*(2), 257-269.

Chadwick, A., Arts, R., Bernstone, C., May, F., Thibeau, S., Zweigel, P. (2008). Best practice for the storage of CO_2 in saline aquifers-observations and guidelines from the SACS and CO2STORE projects (Vol. 14). British Geological Survey.

Chang, C.C., Lin, C.J. (2011). LIBSVM: A library for support vector machines. *ACM Transactions on Intelligent Systems and Technology, 2*(3), 1-27.

Chen, B., Harp, D.R., Lin, Y., Keating, E.H., Pawar, R.J. (2018). Geologic CO_2 sequestration monitoring design: A machine learning and uncertainty quantification based approach. *Applied Energy, 225*, 332-345.

Chen, B., Reynolds, A.C. (2017). Optimal control of ICV's and well operating conditions for the water-alternating-gas injection process. *Journal of Petroleum Science and Engineering, 149*, 623-640.

Chen, C., Chai, Z., Shen, W., Li, W. (2017). Effects of impurities on CO_2 sequestration in saline aquifers: Perspective of interfacial tension and wettability. *Industrial & Engineering Chemistry Research, 57*(1), 371-379.

Chin, L.Y., Raghavan, R., Thomas, L.K. (2000). Fully-coupled geomechanics and fluid-flow analysis of wells with stress-dependent permeability. *SPE Journal, 5*(1), 32-45.

Chong, Z.R., Yang, S.H.B., Babu, P., Linga, P., Li, X.S. (2016). Review of natural gas hydrates as an energy resource: Prospects and challenges. *Applied Energy, 162*, 1633-1652.

Chou, S.I., Vasicek, S.L., Pisio, D.L., Jasek, D.E., Goodgame, J.A. . (1992). CO_2 foam field trial at north ward-estes. In the *67th Annual Technical Conference and Exhibition*, Washington, USA, October 4-7.

Class, H., Ebigbo, A., Helmig, R., Dahle, H.K., Nordbotten, J.M., Celia, M.A., et al. (2009). A benchmark study on problems related to CO_2 storage in geologic formations. *Computational Geosciences, 13*(4), 409-434.

Clavaud, J.B., Maineult, A., Zamora, M., Rasolofosaon, P., Schlitter, C. (2008). Permeability anisotropy and its relations with porous medium structure. *Journal of Geophysical Research, 113*(B1).

Clemens, T., Secklehner, S., Mantatzis, K., Jacobs, B. (2010). Enhanced gas recovery, challenges shown at the example of three gas fields. In *SPE EUROPEC/EAGE Annual Conference and Exhibition*, Barcelona, Spain, June 14-17.

Cooper, C. (2009). A technical basis for carbon dioxide storage. *Energy Procedia, 1*(1), 1727-1733.

Corey, A.T. (1954). The interrelation between gas and oil relative permeabilities. *Producers monthly, 19*(1), 38-41.

Cundall, P. (2008). FLAC 3D Manual: a computer program for fast Lagrangian analysis of continua (Version 4.0). Minneapolis, Minnesota, USA.

Dahaghi, A.K. (2010). Numerical simulation and modeling of enhanced gas recovery and CO_2 sequestration in shale gas reservoirs: A feasibility study. In *SPE International Conference on CO_2 Capture, Storage, and Utilization*, New Orleans, Louisiana, USA, Novemer 10-12.

Dai, Z., Zhang, Y., Bielicki, J., Amooie, M. A., Zhang, M., Yang, C., et al. (2018). Heterogeneity-assisted carbon dioxide storage in marine sediments. *Applied Energy, 225*, 876-883.

Davies, J.P., Davies, D.K. (1999). Stress-dependent permeability: characterization and modeling. In *SPE Annual Technical Conference and Exhibition*, Houston, Texas, UAS, October 3-6.

Davison, J., Freund, P., Smith, A. (2001). Putting carbon back in the ground. Retrieved from https://www.osti.gov/etdeweb/biblio/20204888, accessed on June 22, 2019.

De Marsily, G. (1986). Quantitative hydrogeology; groundwater hydrology for engineers. Academic Press, USA.

De Silva, G.P.D., Ranjith, P.G., Perera, M.S.A. (2015). Geochemical aspects of CO_2 sequestration in deep saline aquifers: A review. *Fuel, 155*, 128-143.

Dempsey, D., Kelkar, S., Pawar, R., Keating, E., Coblentz, D. (2014). Modeling caprock bending stresses and their potential for induced seismicity during CO_2 injection. *International Journal of Greenhouse Gas Control, 22*, 223-236.

Dewers, T., Eichhubl, P., Ganis, B., Gomez, S., Heath, J., Jammoul, M., et al. (2018). Heterogeneity, pore pressure, and injectate chemistry: Control measures for geologic carbon storage. *International Journal of Greenhouse Gas Control, 68*, 203-215.

Doherty, B., Vasylkivska, V., Huerta, N.J., Dilmore, R. (2017). Estimating the leakage along wells during geologic CO_2 storage: Application of the well leakage sssessment tool to a hypothetical storage scenario in Natrona County, Wyoming. *Energy Procedia, 114*, 5151-5172.

Dou, X., Liao, X., Wang, H., Zhao, T., Cheng, Z., Ren, W., Zhang, R. (2015). The study of CO_2 flooding and sequestration in tight gas reservoir: Different completion measures. In *SPE Nigeria Annual International Conference and Exhibition*, Lagos, Nigeria, August 4-6.

Dramsch, J.S., Corte, G., Amini, H., Lüthje, M., MacBeth, C. (2019). Deep learning application for 4D pressure saturation inversion compared to Bayesian inversion on North Sea data. In *Second EAGE Workshop Practical Reservoir Monitoring 2019*, European Association of Geoscientists & Engineers.

Drucker, H., Burges, C.J., Kaufman, L., Smola, A.J., Vapnik, V. (1997). Support Vector Regression Machines. In *Advances in neural information processing systems*, 155-161.

Ebigbo, A., Class, H., Helmig, R. (2006). CO_2 leakage through an abandoned well: problem-oriented benchmarks. *Computational Geosciences, 11*(2), 103-115.

Edenhofer, O., Pichs-Madruga, R., Sokona, Y., Kadner, S., Minx, J., Brunner, S. (2014). Change 2014: Mitigation of climate change. Contribution of Working Group III to the Fifth Assessment Report of the Intergovernmental Panel on Climate Change.

Eiken, O., Ringrose, P., Hermanrud, C., Nazarian, B., Torp, T.A., Høier, L. (2011). Lessons learned from 14 years of CCS operations: Sleipner, In Salah and Snøhvit. *Energy Procedia, 4*, 5541-5548.

Eliebid, M., Mahmoud, M., Shawabkeh, R., Elkatatny, S., Hussein, I.A. (2018). Effect of CO_2 adsorption on enhanced natural gas recovery and sequestration in carbonate reservoirs. *Journal Of Natural Gas Science And Engineering, 55*, 575-584.

Etheridge, D., Luhar, A., Loh, Z., Leuning, R., Spencer, D., Steele, P., et al. (2011). Atmospheric monitoring of the CO2CRC Otway Project and lessons for large scale CO_2 storage projects. *Energy Procedia, 4*, 3666-3675.

Evans, D.J., Chadwick, R.A. (2009). Underground gas storage: Worldwide experiences and future development in the UK and Europe. Geological Society of London.

Fang, Z., Li, X., Hu, H. (2011). Gas mixture enhance coalbed methane recovery technology: Pilot tests. *Energy Procedia, 4*, 2144-2149.

Feather, B., Archer, R. (2010). Enhanced natural gas recovery by carbon dioxide injection for storage puposes. In the *17th Australasian Fluid Mechanics Conference*, Auckland, New Zealand, December 5-9.

Fei, W.B., Li, Q., Liu, X.H., Wei, X.C., Jing, M., Song, R.R., et al. (2014). Coupled analysis for interaction of coal mining and CO 2 geological storage in Ordos Basin, China. In *the 8th Asian Rock Mechanics Symposium*, October 14-16.

Fer, I., Haugan, P.M. (2003). Dissolution from a liquid CO_2 lake disposed in the deep ocean. *Limnology and Oceanography, 48*(2), 872-883.

Finley, R.J., Frailey, S.M., Leetaru, H.E., Senel, O., Couëslan, M.L., Scott, M. (2013). Early operational experience at a one-million tonne CCS demonstration project, Decatur, Illinois, USA. *Energy Procedia, 37*, 6149-6155.

Fischer, S., Wolf, L., Fuhrmann, L., Gahre, H., Rütters, H. . (2018). Simulated fluid-rock interactions during storage of temporally varying impure CO_2 streams. In the *Fifth CO_2 Geological Storage Workshop*, Utrecht, The Netherlands, November 21-23.

Flett, M.A., Beacher, G.J., Brantjes, J., Burt, A.J., Dauth, C., Koelmeyer, F.M., et al. (2008). Gorgon project: subsurface evaluation of carbon dioxide disposal under Barrow Island. In *SPE Asia Pacific Oil and Gas Conference and Exhibition*, Perth, Australia, October 20-22.

Gaus, I. (2010). Role and impact of CO_2–rock interactions during CO_2 storage in sedimentary rocks. *International Journal of Greenhouse Gas Control, 4*(1), 73-89.

Gaus, I., Audigane, P., André, L., Lions, J., Jacquemet, N., Durst, P., et al. (2008). Geochemical and solute transport modelling for CO_2 storage, what to expect from it? *International Journal of Greenhouse Gas Control, 2*(4), 605-625.

Gawel, K., Todorovic, J., Liebscher, A., Wiese, B., Opedal, N. (2017). Study of materials retrieved from a Ketzin CO_2 monitoring well. *Energy Procedia, 114*, 5799-5815.

GCCSI. (2011). Economic assessment of carbon capture and storage technologies: 2011 Update. Retrieved from www.globalccsinstitute.com/publications/economicassessment-carbon-capture-and-storage-technologies-2011-update, accessed on February 3, 2019.

Geel, C.R., Arts, R.J., Van Eijs, R.M.H.E., Kreft, E., Hartman, J., D'Hoore, D. (2006). Geological site characterization of the nearly depleted K12-B gas field, offshore the Netherlands. In Procedings, CO2SC Symposium, Lawrence Berkeley National Laboratory, Berkeley, California, March 20-22.

Gemmer, L., Hansen, O., Iding, M., Leary, S., Ringrose, P. (2012). Geomechanical response to CO_2 injection at Krechba, In salah, Algeria. *First Break, 30*(2), 79-84.

Gercek, H. (2007). Poisson's ratio values for rocks. *International Journal of Rock Mechanics and Mining Sciences, 44*(1), 1-13.

Gilfillan, S.M., Lollar, B.S., Holland, G., Blagburn, D., Stevens, S., Schoell, M., et al. (2009). Solubility trapping in formation water as dominant CO_2 sink in natural gas fields. *Nature, 458*(7238), 614-618.

Gislason, S.R., Oelkers, E.H. (2014). Carbon storage in basalt. *Science, 344*(6182), 373-374.

Godec, M., Koperna, G., Gale, J. (2014). CO_2-ECBM: A review of its status and global potential. *Energy Procedia, 63*, 5858-5869.

Goldberg, D.S., Takahashi, T., Slagle, A.L. (2008). Carbon dioxide sequestration in deep-sea basalt. *Proceedings of the National Academy of Sciences, 105*(29), 9920-9925.

Götz, J., Lüth, S., Henninges, J., Reinsch, T. (2018). Vertical seismic profiling using a daisy-chained deployment of fibre-optic cables in four wells simultaneously - Case study at the Ketzin carbon dioxide storage site. *Geophysical Prospecting, 66*(6), 1201-1214.

Gou, Y., Hou, Z., Li, M., Feng, W., Liu, H. (2016). Coupled thermo–hydro–mechanical simulation of CO_2 enhanced gas recovery with an extended equation of state module for TOUGH2MP-FLAC3D. *Journal of Rock Mechanics and Geotechnical Engineering, 8*(6), 904-920.

Gou, Y., Hou, Z., Liu, H., Zhou, L., Were, P. (2014). Numerical simulation of carbon dioxide injection for enhanced gas recovery (CO_2-EGR) in Altmark natural gas field. *Acta Geotechnica, 9*(1), 49-58.

Gozalpour, F., Ren, S.R., Tohidi, B. (2006). CO_2 EOR and storage in oil reservoir. *Oil & gas science and technology, 60*(3), 537-546.

Guo, F., Aryana, S.A. (2018). Improved sweep efficiency due to foam flooding in a heterogeneous microfluidic device. *Journal of Petroleum Science and Engineering, 164*, 155-163.

Guo, F., Aryana, S.A., Wang, Y., McLaughlin, J.F., Coddington, K. (2019). Enhancement of storage capacity of CO_2 in megaporous saline aquifers using nanoparticle-stabilized CO_2 foam. *International Journal of Greenhouse Gas Control, 87*, 134-141.

Hajiw, M, Corvisier, J, El Ahmar, E, Coquelet, C. (2018). Impact of impurities on CO_2 storage in saline aquifers: Modelling of gases solubility in water. *International Journal of Greenhouse Gas Control, 68*, 247-255.

Hannis, S., Lu, J., Chadwick, A., Hovorka, S., Kirk, K., Romanak, K., et al. (2017). CO_2 storage in depleted or depleting oil and gas fields: What can we learn from existing projects? *Energy Procedia, 114*, 5680-5690.

Hansen, O., Gilding, D., Nazarian, B., Osdal, B., Ringrose, P., Kristoffersen, J.B., et al. (2013). Snøhvit: The history of injecting and storing 1 Mt CO_2 in the fluvial Tubåen Fm. *Energy Procedia, 37*, 3565-3573.

Harris, J., Kovscek, A.R., Orr, F.M., Zoback, M.D. (2009). Geologic storage of CO_2 in coal beds. Global Climate and Energy Project (GCEP) Technical Report, Stanford University, Stanford, California, USA.

Hart, D.J., Wang, H.F. (1995). Laboratory measurements of a complete set of poroelastic moduli for Berea sandstone and Indiana limestone. *Journal of Geophysical Research: Solid Earth, 100*(B9), 17741-17751.

Hattenbach, R.P., Wilson, M., Brown, K.R. . (1998). Capture of carbon dioxide from coal combustion and its utilization for enhanced oil recovery. In GHGT-4 Conf, United Kingdom.

Hawcroft, M., Walsh, E., Hodges, K., Zappa, G. (2018). Significantly increased extreme precipitation expected in Europe and North America from extratropical cyclones. *Environmental Research Letters, 13*, 124006.

Hawthorne, S.B., Gorecki, C.D., Sorensen, J.A., Steadman, E.N., Harju, J.A., Melzer, S. (2013). Hydrocarbon mobilization mechanisms from upper, middle, and lower bakken reservoir rocks exposed to CO_2. In *SPE Unconventional Resources Conference-Canada*. Calgary, Alberta, Canada, November 5-7.

Homma, T., Saltelli, A. (1996). Importance measures in global sensitivity analysis of nonlinear models. *Reliability Engineering & System Safety, 52*(1), 1-17.

Honari, A., Bijeljic, B., Johns, M.L., May, E.F. (2015). Enhanced gas recovery with CO_2 sequestration: The effect of medium heterogeneity on the dispersion of supercritical CO_2-CH_4. *International Journal of Greenhouse Gas Control, 39*, 39-50.

Honari, A., Zecca, M., Vogt, S.J., Iglauer, S., Bijeljic, B., Johns, M.L., et al. (2016). The impact of residual water on CH_4-CO_2 dispersion in consolidated rock cores. *International Journal of Greenhouse Gas Control, 50*, 100-111.

Hou, Z., Bacon, D.H., Engel, D.W., Lin, G., Fang, Y., Ren, H., et al. (2014). Uncertainty analyses of CO_2 plume expansion subsequent to wellbore CO_2 leakage into aquifers. *International Journal of Greenhouse Gas Control, 27*, 69-80.

Hou, Z., Tadongmo, F.A., Pusch, G. (2009). Secondary in situ stresses resulting from change of pore pressure influence on the maximal storage pressure of CO_2 or natural gas in reservoirs. In *DGMK/ÖGEW-Frühjahrstagung band*, Celle, Germany.

House, K.Z., Schrag, D.P., Harvey, C.F., Lackner, K.S. . (2006). Permanent carbon dioxide storage in deep-sea sediments. *Proceedings of the National Academy of Sciences, 103*(33), 12291-12295.

Huang, X., Bandilla, K.W., Celia, M.A., Bachu, S. (2014). Basin-scale modeling of CO_2 storage using models of varying complexity. *International Journal of Greenhouse Gas Control, 20*, 73-86.

IEA. (2009). Technology roadmap carbon capture and storage. Retrieved from https://insideclimatenews.org/sites/default/files/IEA-CCS%20Roadmap.pdf, accessed on June 13, 2019.

IEA. (2010). Energy technology perspectives, scenarios and strategies to 2050. Retrieved from https://www.iea.org/publications/freepublications/publication/etp2010.pdf, accessed on June 9, 2019.

IEF. (2020). IEF insight brief - the circular carbon economy. Retrieved from https://www.ief.org/_resources/files/comparative-analysis/march-ief-insight-brief---the-circular-carbon-economy.pdf, accessed on October 9, 2020.

Imre, A.R., Groniewsky, A., Györke, G., Katona, A., Velmovszki, D. (2019). Anomalous properties of some fluids – with high relevance in energy engineering – in their pseudo-critical (Widom) region. *Periodica Polytechnica Chemical Engineering, 63*(2), 276-285.

Imre, A.R., Ramboz, C., Deiters, U.K., Kraska, T. (2014). Anomalous fluid properties of carbon dioxide in the supercritical region: application to geological CO_2 storage and related hazards. *Environmental Earth Sciences, 73*(8), 4373-4384.

Global CCS Institute. (2019). Global CCS facilities database. Retrieved from https://co2re.co/FacilityData, accessed on November 15, 2019.

Iogna, A., Guillet-Lhermite, J., Wood, C., Deflandre, J.P. (2017). CO_2 storage and enhanced gas recovery: Using extended black oil modelling to simulate CO_2 injection on a North Sea depleted gas field. In *SPE Europec featured at 79th EAGE Conference and Exhibition*, Paris, France, June 12-15.

Irlam, L. (2017). Global costs of carbon capture and storage - 2017 Update. Retrieved from https://www.globalccsinstitute.com/archive/hub/publications/201688/global-ccs-cost-updatev4.pdf, accessed on July 7, 2020.

Itasca. (2009). Fast Lagrangian analysis of continua in 3 dimensions, Version 4.0. Itasca Consulting Group, Minneapolis, Minnesota, USA.

Jafari Raad, S.M., Hassanzadeh, H. (2017). Prospect for storage of impure carbon dioxide streams in deep saline aquifers—A convective dissolution perspective. *International Journal of Greenhouse Gas Control, 63*, 350-355.

Jahangiri, H.R., Zhang, D. (2012). Ensemble based co-optimization of carbon dioxide sequestration and enhanced oil recovery. *International Journal of Greenhouse Gas Control, 8*, 22-33.

Jayne, R.S., Wu, H., Pollyea, R.M. (2018). Geologic CO_2 sequestration in a Basalt reservoir: Constraining permeability uncertainty within the Columbia River Basalt Group. In *Proceedings of TOUGH Symposium 2018*, Lawrence Berkeley National Laboratory, Berkeley, California, October 8-10.

Jayne, R.S., Wu, H., Pollyea, R.M. (2019). Geologic CO_2 sequestration and permeability uncertainty in a highly heterogeneous reservoir. *International Journal of Greenhouse Gas Control, 83*, 128-139.

Jenkins, C.R., Cook, P.J., Ennis-King, J., Undershultz, J., Boreham, C., Dance, T., et al. (2012). Safe storage and effective monitoring of CO_2 in depleted gas fields. *Proceedings of the National Academy of Sciences, 109*(2), E35-41.

Jeong, H., Srinivasan, S. (2016). Fast assessment of CO_2 plume characteristics using a connectivity based proxy. *International Journal of Greenhouse Gas Control, 49*, 387-412.

Jia, B., Tsau, J.S., Barati, R. (2019). A review of the current progress of CO_2 injection EOR and carbon storage in shale oil reservoirs. *Fuel, 236*, 404-427.

Jia, W., McPherson, B.J., Pan, F., Xiao, T., Bromhal, G. (2016). Probabilistic analysis of CO_2 storage mechanisms in a CO_2-EOR field using polynomial chaos expansion. *International Journal of Greenhouse Gas Control, 51*, 218-229.

Jikich, S.A., Smith, D.H., Sams, W.N., Bromhal, G.S. (2003). Enhanced Gas Recovery (EGR) with carbon dioxide sequestration: A simulation study of effects of injection strategy and operational parameters. In SPE Eastern Regional/AAPG Eastern Section Joint Meeting, Pittsburgh, Pennsylvania, USA, September 6-10.

Jin, X., Shah, S.N., Roegiers, J.C., Zhang, B. (2014). Fracability evaluation in shale reservoirs-an integrated petrophysics and geomechanics approach. In *SPE hydraulic fracturing technology conference*, The Woodlands, Texas, USA, February 4-6.

Johnston, K.P., Da Rocha, S.R.P. (2009). Colloids in supercritical fluids over the last 20 years and future directions. *The Journal of Supercritical Fluids, 47*(3), 523-530.

Juez-Larré, J., Remmelts, G., Breunese, J.N., van Gessel, S.F., Leeuwenburgh, O. (2016). Using underground gas storage to replace the swing capacity of the giant natural gas field of Groningen in the Netherlands. A reservoir performance feasibility study. *Journal of Petroleum Science and Engineering, 145*, 34-53.

Jung, H., Singh, G., Espinoza, D.N., Wheeler, M.F. (2018). Quantification of a maximum injection volume of CO_2 to avert geomechanical perturbations using a compositional fluid flow reservoir simulator. *Advances in Water Resources, 112*, 160-169.

Jung, H.B., Um, W., Cantrell, K.J. (2013). Effect of oxygen co-injected with carbon dioxide on Gothic shale caprock–CO_2–brine interaction during geologic carbon sequestration. *Chemical Geology, 354*, 1-14.

Kabirzadeh, H., Sideris, M.G., Shin, Y.J., Kim, J.W. (2017). Gravimetric monitoring of confined and unconfined geological CO_2 reservoirs. *Energy Procedia, 114*, 3961-3968.

Kane, A.V. (1979). Performance review of a large-scale CO_2-WAG enhanced recovery project, SACROC Unit Kelly-Snyder Field. *Journal of Petroleum Technology, 31*(2), 217-231.

Katz, D.L., Tek, M.R. (1981). Overview on underground storage of natural gas. *Journal of Petroleum Technology, 33*(06), 943-951.

Kelemen, P.B., Matter, J. (2008). In situ carbonation of peridotite for CO_2 storage. *Proceedings of the National Academy of Sciences, 105*(45), 17295–17300.

Kemper, J. (2015). Biomass and carbon dioxide capture and storage: A review. *International Journal of Greenhouse Gas Control, 40*, 401-430.

Khan, C., Amin, R., Madden, G. (2012). Economic modelling of CO_2 injection for enhanced gas recovery and storage: A reservoir simulation study of operational parameters. *Energy and Environment Research, 2*(2), 65-82.

Khan, C., Amin, R., Madden, G. (2013). Carbon dioxide injection for enhanced gas recovery and storage (reservoir simulation). *Egyptian Journal of Petroleum, 22*(2), 225-240.

Khan, C.M.H. (2013). Techno-economic reservoir simulation model for CO_2 sequestration evaluation. Doctoral dissertation, Curtin University.

Kim, J., Choi, J., Park, K. (2015). Comparison of nitrogen and carbon dioxide as cushion gas for underground gas storage reservoir. *Geosystem Engineering, 18*(3), 163-167.

Klimkowski, Ł., Nagy, S., Papiernik, B., Orlic, B., Kempka, T. (2015). Numerical simulations of enhanced gas recovery at the Załęcze gas field in Poland confirm high CO_2 storage capacity and mechanical integrity. *Oil & Gas Science and Technology – Revue d'IFP Energies nouvelles, 70*(4), 655-680.

Klinkenberg, L.J. (1941). The permeability of porous media to liquids and gases. *Drilling and Production Practice*, 200-213.

Kobos, P.H., Cappelle, M.A., Krumhansl, J.L., Dewers, T.A., McNemar, A., Borns, D.J. (2011). Combining power plant water needs and carbon dioxide storage using saline formations: Implications for carbon dioxide and water management policies. *International Journal of Greenhouse Gas Control, 5*(4), 899-910.

Kodama, S., Nishimoto, T., Yamamoto, N., Yogo, K., Yamada, K. (2008). Development of a new pH-swing CO_2 mineralization process with a recyclable reaction solution. *Energy, 33*(5), 776-784.

Koide, H., Shindo, Y., Tazaki, Y., Iijima, M., Ito, K., Kimura, N., et al. (1997). Deep sub-seabed disposal of CO_2—The most protective storage. *Energy Conversion and Management, 38*, S253-S258.

Koytsoumpa, E.I., Bergins, C., Kakaras, E. (2018). The CO_2 economy: Review of CO_2 capture and reuse technologies. *The Journal of Supercritical Fluids, 132*, 3-16.

Kreft, E., Bernstone, C., Meyer, R., May, F., Arts, R., Obdam, A., et al. (2007). "The Schweinrich structure", a potential site for industrial scale CO_2 storage and a test case for safety assessment in Germany. *International Journal of Greenhouse Gas Control, 1*(1), 69-74.

Krevor, S., Blunt, M.J., Benson, S.M., Pentland, C.H., Reynolds, C., Al-Menhali, A., et al. (2015). Capillary trapping for geologic carbon dioxide storage – From pore scale physics to field scale implications. *International Journal of Greenhouse Gas Control, 40*, 221-237.

Kühn, M., Tesmer, M., Pilz, P., Meyer, R., Reinicke, K., Förster, A., et al. (2012). CLEAN: project overview on CO_2 large-scale enhanced gas recovery in the Altmark natural gas field (Germany). *Environmental Earth Sciences, 67*(2), 311-321.

Laherrère, J. (1997). Distribution and evolution of "recovery factor". In *Oil Reserves Conference*, Paris, France, November 11.

Laier, T. (2012). Results of monitoring groundwater above the natural gas underground storage at Stenlille, Denmark. *Geological Survey of Denmark and Greenland Bulletin, 26*, 45-48.

Langston, M.V., Hoadley, S.F., Young, D.N. (1988). Definitive CO_2 flooding response in the SACROC unit. In *SPE/DOE Enhanced Oil Recovery Symposium*, Tulsa, Oklahoma, UAS, April 17-20.

Van, S.L., Chon, B.H. (2017). Evaluating the critical performances of a CO_2–Enhanced oil recovery process using artificial neural network models. *Journal of Petroleum Science and Engineering, 157*, 207-222.

Lee, J.H., Park, Y.C., Sung, W.M., Lee, Y.S. (2010). A simulation of a trap mechanism for the sequestration of CO_2 into Gorae V Aquifer, Korea. *Energy Sources, Part A: Recovery, Utilization, and Environmental Effects, 32*(9), 796-808.

Lee, Y., Choi, W., Shin, K., Seo, Y. (2017). CH_4 -CO_2 replacement occurring in sII natural gas hydrates for CH_4 recovery and CO_2 sequestration. *Energy Conversion and Management, 150*, 356-364.

Lei, H., Li, J., Li, X., Jiang, Z. (2016). Numerical modeling of co-injection of N_2 and O_2 with CO_2 into aquifers at the Tongliao CCS site. *International Journal of Greenhouse Gas Control, 54*, 228-241.

Lengler, U., De Lucia, M., Kühn, M. (2010). The impact of heterogeneity on the distribution of CO_2: Numerical simulation of CO_2 storage at Ketzin. *International Journal of Greenhouse Gas Control, 4*(6), 1016-1025.

Leung, D.Y.C., Caramanna, G., Maroto-Valer, M.M. (2014). An overview of current status of carbon dioxide capture and storage technologies. *Renewable and Sustainable Energy Reviews, 39*, 426-443.

Levine, J.S., Matter, J.M., Goldberg, D., Cook, A., Lackner, K.S. (2007). Gravitational trapping of carbon dioxide in deep sea sediments: Permeability, buoyancy, and geomechanical analysis. *Geophysical Research Letters, 34*(24).

Li, C., Laloui, L. (2017). Coupled thermo-hydro-mechanical effects on caprock stability during carbon dioxide injection. *Energy Procedia, 114*, 3202-3209.

Li, D., Jiang, X. (2014). A numerical study of the impurity effects of nitrogen and sulfur dioxide on the solubility trapping of carbon dioxide geological storage. *Applied Energy, 128*, 60-74.

Li, D., Jiang, X. (2017). Numerical investigation of the partitioning phenomenon of carbon dioxide and multiple impurities in deep saline aquifers. *Applied Energy, 185*, 1411-1423.

Li, D., Jiang, X. (2020). Numerical investigation of convective mixing in impure CO_2 geological storage into deep saline aquifers. *International Journal of Greenhouse Gas Control, 96*, 103015.

Li, D., Jiang, X., Meng, Q., Xie, Q. (2015). Numerical analyses of the effects of nitrogen on the dissolution trapping mechanism of carbon dioxide geological storage. *Computers & Fluids, 114*, 1-11.

Li, H., Jakobsen, J.P., Wilhelmsen, Ø., Yan, J. (2011). PVTxy properties of CO_2 mixtures relevant for CO_2 capture, transport and storage: Review of available experimental data and theoretical models. *Applied Energy, 88*(11), 3567-3579.

Li, J., Mei, K., Zhang, M. (2012). Global parameter sensitivity analysis on water quality model. *Acta Agriculturae Zhejiangensis, 24*(2), 314-320.

Li, L., Zhao, N., Wei, W., Sun, Y. (2013). A review of research progress on CO_2 capture, storage, and utilization in Chinese Academy of Sciences. *Fuel, 108*, 112-130.

Li, P., Zhou, D., Liang, X. (2018). Geological characterization and simulation of an offshore oilfield (HZ21-1) for CO_2 storage in the Pearl River Mouth Basin. In *CHINA ROCK 2018-The 15th China Rock Mechanics and Engineering Academic Annual Meeting*, Beijing, China, Semptember 19-22.

Li, Q., Liu, G. (2016). Risk assessment of the geological storage of CO_2: A Review. Geologic Carbon Sequestration, Springer International Publishing Switzerland, 249-284.

Li, Q., Wei, Y.N., Liu, G., Shi, H. (2015). CO_2-EWR: a cleaner solution for coal chemical industry in China. *Journal of Cleaner Production, 103*, 330-337.

Liebscher, A., Möller, F., Bannach, A., Köhler, S., Wiebach, J., Schmidt-Hattenberger, C., et al. (2013). Injection operation and operational pressure–temperature monitoring at the CO_2 storage pilot site Ketzin, Germany—Design, results, recommendations. *International Journal of Greenhouse Gas Control, 15*, 163-173.

Liu, F., Ellett, K., Xiao, Y., Rupp, J.A. (2013). Assessing the feasibility of CO_2 storage in the New Albany Shale (Devonian–Mississippian) with potential enhanced gas recovery using reservoir simulation. *International Journal of Greenhouse Gas Control, 17*, 111-126.

Liu, G., Gorecki, C.D., Saini, D., Bremer, J.M., Klapperich, R.J., Braunberger, J.R. (2013). Four-site case study of water extraction from CO_2 storage reservoirs. *Energy Procedia, 37*, 4518-4525.

Liu, H. (2014). Numerical study of physico-chemical interactions for CO_2 sequestration and geothermal energy utilization in the Ordos Basin, China. Doctoral dissertation, Clausthal University of Technology.

Liu, H., Hou, Z., Li, X., Wei, N., Tan, X., Were, P. (2015). A preliminary site selection system for a CO_2-AGES project and its application in China. *Environmental Earth Sciences, 73*(11), 6855-6870.

Liu, H., Hou, Z., Were, P., Sun, X., Gou, Y. (2015). Numerical studies on CO_2 injection–brine extraction process in a low-medium temperature reservoir system. *Environmental Earth Sciences, 73*(11), 6839-6854.

Liu, H.H., Houseworth, J., Rutqvist, J., Zheng, L., Asahina, D., Li, L., et al. (2013). Report on THMC modeling of the near field evolution of a generic clay repository: Model validation and demonstration. Berkeley, California, USA.

Liu, H.J., Were, P., Li, Q., Gou, Y., Hou, Z. (2017). Worldwide status of CCUS technologies and their development and challenges in China. *Geofluids, 2017*, 1-25.

Liu, K., Yu, Z., Saeedi, A., Esteban, L. (2015). Effects of permeability, heterogeneity and gravity on supercritical CO_2 displacing gas under reservoir conditions. In *SPE Enhanced Oil Recovery Conference*, Kuala Lumpur, Malaysia, August 11-13.

Liu, S., Song, Y., Zhao, C., Zhang, Y., Lv, P., Jiang, L., et al. (2018). The horizontal dispersion properties of CO_2-CH_4 in sand packs with CO_2 displacing the simulated natural gas. *Journal Of Natural Gas Science And Engineering, 50*, 293-300.

Liu, S., Zhang, Y., Zhao, J., Jiang, L., Song, Y. (2020). Dispersion characteristics of CO_2 enhanced gas recovery over a wide range of temperature and pressure. *Journal of Natural Gas Science and Engineering, 73*, 103056.

Liu, Y., Hou, J., Zhao, H., Liu, X., Xia, Z. (2018). A method to recover natural gas hydrates with geothermal energy conveyed by CO_2. *Energy, 144*, 265-278.

Loeve, D., Hofstee, C., Maas, J.G. (2014). Thermal effects in a depleted gas field by cold CO_2 injection in the presence of methane. *Energy Procedia, 63*, 5378-5393.

Lui, L.C., Leamon, G. (2014). Developments towards environmental regulation of CCUS projects in China. *Energy Procedia, 63*, 6903-6911.

Lumley, D. (2010). 4D seismic monitoring of CO_2 sequestration. In *ASEG 2010,* Sydney, Australia.

Luo, F., Xu, R.N., Jiang, P.X. (2013). Numerical investigation of the influence of vertical permeability heterogeneity in stratified formation and of injection/production well perforation placement on CO_2 geological storage with enhanced CH_4 recovery. *Applied Energy, 102*, 1314-1323.

Lv, G., Li, Q., Wang, S., Li, X. (2015). Key techniques of reservoir engineering and injection–production process for CO_2 flooding in China's SINOPEC Shengli Oilfield. *Journal of CO_2 Utilization, 11*, 31-40.

Ma, J., Li, Q., Kempka, T., Kuehn, M. (2019). Hydromechanical response and impact of gas mixing behavior in subsurface CH_4 storage with CO_2-based cushion gas. *Energy & Fuels, 33*, 6527–6541.

Mac Dowell, N., Fennell, P.S., Shah, N., Maitland, G.C. (2017). The role of CO_2 capture and utilization in mitigating climate change. *Nature Climate Change, 7*(4), 243-249.

MacDowell, N., Florin, N., Buchard, A., Hallett, J., Galindo, A., Jackson, G., et al. (2010). An overview of CO_2 capture technologies. *Energy & Environmental Science, 3*(11), 1645-1669.

Mahmoodpour, S., Rostami, B., Emami-Meybodi, H. (2018). Onset of convection controlled by N_2 impurity during CO_2 storage in saline aquifers. *International Journal of Greenhouse Gas Control, 79*, 234-247.

Mamora, D.D., Seo, J.G. (2002). Enhanced gas recovery by carbon dioxide sequestration in depleted gas reservoirs. In *SPE Annual Technical Conference and Exhibition*, San Antonio, Texas, USA, September 29-October 2.

Marston, P.M. (2017). Incidentally speaking: A systematic assessment and comparison of incidental storage of CO_2 during EOR with other near-term storage options. *Energy Procedia, 114*, 7422-7430.

Martens, S., Kempka, T., Liebscher, A., Lüth, S., Möller, F., Myrttinen, A., et al. (2012). Europe's longest-operating on-shore CO_2 storage site at Ketzin, Germany: a progress report after three years of injection. *Environmental Earth Sciences, 67*(2), 323-334.

Mathias, S.A., Gluyas, J.G., Oldenburg, C.M., Tsang, C.F. (2010). Analytical solution for Joule–Thomson cooling during CO_2 geo-sequestration in depleted oil and gas reservoirs. *International Journal of Greenhouse Gas Control, 4*(5), 806-810.

Mathisen, A., Skagestad, R. (2017). Utilization of CO_2 from Emitters in Poland for CO_2-EOR. *Energy Procedia, 114*, 6721-6729.

Matter, J.M., Stute, M., Snæbjörnsdottir, S.Ó., Oelkers, E.H., Gislason, S.R., Aradottir, E.S., et al. (2016). Rapid carbon mineralization for permanent disposal of anthropogenic carbon dioxide emissions. *Science, 352*(6291), 1312-1314.

Mawalkar, S., Brock, D., Burchwell, A., Kelley, M., Mishra, S., Gupta, N., et al. (2019). Where is that CO_2 flowing? Using Distributed Temperature Sensing (DTS) technology for monitoring injection of CO_2 into a depleted oil reservoir. *International Journal of Greenhouse Gas Control, 85*, 132-142.

Mayer, B., Humez, P., Becker, V., Dalkhaa, C., Rock, L., Myrttinen, A., et al. (2015). Assessing the usefulness of the isotopic composition of CO_2 for leakage monitoring at CO_2 storage sites: A review. *International Journal of Greenhouse Gas Control, 37*, 46-60.

Metcalfe, R.S. (1982). Effects of impurities on minimum miscibility pressures and minimum enrichment levels for CO_2 and rich-gas displacements. *Society of Petroleum Engineers Journal, 22*(2), 219-225.

Metz, B., Davidson, O., De Coninck, H., Loos, M., Meyer, L. (2005). Carbon dioxide capture and storage. Special Report of the Intergovernmental Panel on Climate Change. Cambridge University Press, UK.

Michael, K., Golab, A., Shulakova, V., Ennis-King, J., Allinson, G., Sharma, S., et al. (2010). Geological storage of CO_2 in saline aquifers—A review of the experience from existing storage operations. *International Journal of Greenhouse Gas Control, 4*(4), 659-667.

Mishra, S., Ganesh, P.R., Schuetter, J. (2017). Developing and validating simplified predictive models for CO_2 geologic sequestration. *Energy Procedia, 114*, 3456-3464.

Misra, B.R., Foh, S.E., Shikari, Y.A., Berry, R.M., Labaune, F. (1988). The use of inert base gas in underground natural gas storage. In *SPE Gas Technology Symposium*, Dallas, Texas, USA, June 13-15.

Morozova, D., Zettlitzer, M., Let, D., Würdemann, H. (2011). Monitoring of the microbial community composition in deep subsurface saline aquifers during CO_2 storage in Ketzin, Germany. *Energy Procedia, 4*, 4362-4370.

Namhata, A., Oladyshkin, S., Dilmore, R.M., Zhang, L., Nakles, D.V. (2016). Probabilistic assessment of above zone pressure predictions at a geologic carbon storage site. *Scientific reports, 6*, 39536.

Namhata, A., Zhang, L., Dilmore, R.M., Oladyshkin, S., Nakles, D.V. (2017). Modeling changes in pressure due to migration of fluids into the Above Zone Monitoring Interval of a geologic carbon storage site. *International Journal of Greenhouse Gas Control, 56*, 30-42.

Narinesingh, J., Alexander, D. (2016). Injection well placement analysis for optimizing CO_2 enhanced gas recovery coupled with sequestration in condensate reservoirs. In *SPE Trinidad and Tobago Section Energy Resources Conference*, Port of Spain, Trinidad and Tobago, June 13-15.

Narinesingh, J., Alexander, D. (2014). CO_2 enhanced gas recovery and geologic sequestration in condensate reservoir: A simulation study of the effects of injection pressure on condensate recovery from reservoir and CO_2 storage efficiency. *Energy Procedia, 63*, 3107-3115.

National standards of the P.R. China. (2018) Natural Gas, GB17820—2018.

NETL. (2017). U.S. DOE's National Risk Assessment Partnership: Assessing carbon storage risk performance to support decision making amidst uncertainty. Retrieved from https://www.cslforum.org/cslf/sites/default/files/documents/AbuDhabi2017/Bromhal-NRAP-TG-AbuDhabi0517.pdf, accessed on May 12, 2019.

Newell, P., Yoon, H., Martinez, M.J., Bishop, J.E., Bryant, S.L. (2017). Investigation of the influence of geomechanical and hydrogeological properties on surface uplift at In Salah. *Journal of Petroleum Science and Engineering, 155*, 34-45.

Nicot, J.P., Solano, S., Lu, J., Mickler, P., Romanak, K., Yang, C., Zhang, X. (2013). Potential subsurface impacts of CO_2 stream impurities on geologic carbon storage. *Energy Procedia, 37*, 4552-4559.

Nihan Çetin, D., Demirel, T., Deveci, M., Vardar, G. (2017). Location selection for underground natural gas storage using Choquet integral. *Journal Of Natural Gas Science And Engineering, 45*, 368-379.

Niu, C., Tan, Y. (2014). Numerical simulation and analysis of migration law of gas mixture using carbon dioxide as cushion gas in underground gas storage reservoir. *Journal of Harbin Institute of Technology (New Series), 21*(3), 121-128.

Nordbotten, J.M., Celia, M.A., Bachu, S. (2005). Injection and storage of CO_2 in deep saline aquifers: Analytical solution for CO_2 plume evolution during injection. *Transport in Porous Media, 58*(3), 339-360.

Nordbotten, J.M., Flemisch, B., Gasda, S.E., Nilsen, H.M., Fan, Y., Pickup, G.E., et al. (2012). Uncertainties in practical simulation of CO_2 storage. *International Journal of Greenhouse Gas Control, 9*, 234-242.

O'Connor, W.K., Dahlin, D.C., Nilsen, D.N., Gerdemann, S.J., Rush, G.E., Walters, R.P., et al. (2001). Research status on the sequestration of carbon dioxide by direct aqueous mineral carbonation. In the *18th Annual International Pittsburgh Coal Conference*, Newcastle, NSW, Australia, December 3-7.

Odi, U. (2012). Analysis and potential of CO_2 huff-n-puff for near wellbore condensate removal and enhanced gas recovery. In *SPE Annual Technical Conference and Exhibition*, San Antonio, Texas, USA, October 8-10.

Odi, U. (2013). Optimal process design for coupled CO_2 sequestration and enhanced gas recovery in carbonate reservoirs. Doctoral dissertation, Texas A&M University,

Met Office. (2017). Our changing world - global indicators. Retrieved from https://www.metoffice.gov.uk/climate-guide, accessed on June 9, 2019.

Oh, T.H. (2010). Carbon capture and storage potential in coal-fired plant in Malaysia—A review. *Renewable and Sustainable Energy Reviews, 14*(9), 2697-2709.

Oldenburg, C., Pruess, K., Benson, S.M. (2001). Process modeling of CO_2 injection into natural gas reservoirs for carbon sequestration and enhanced gas recovery. *Energy & Fuels, 15*(2), 293-298.

Oldenburg, C.M. (2003). Carbon dioxide as cushion gas for natural gas storage. *Energy & Fuels, 17*(1), 240-246.

Oldenburg, C.M. (2007). Joule-Thomson cooling due to CO_2 injection into natural gas reservoirs. *Energy Conversion and Management, 48*(6), 1808-1815.

Oldenburg, C.M., Pan, L. (2013). Utilization of CO_2 as cushion gas for porous media compressed air energy storage. *Greenhouse Gases: Science and Technology, 3*(2), 124-135.

Onuma, T., Okada, K., Otsubo, A. (2011). Time series analysis of surface deformation related with CO_2 injection by satellite-borne SAR interferometry at In Salah, Algeria. *Energy Procedia, 4*, 3428-3434.

Opedal, N. (2018). Ensuring integrity of CO_2 storage: An overview of ongoing experimental activity. In the Fifth CO_2 Geological Storage Workshop, Utrecht, The Netherlands, November 21-23.

Orlic, B. (2016). Geomechanical effects of CO_2 storage in depleted gas reservoirs in the Netherlands: Inferences from feasibility studies and comparison with aquifer storage. *Journal of Rock Mechanics and Geotechnical Engineering, 8*(6), 846-859.

Ota, M., Abe, Y., Watanabe, M., Smith, R.L., Inomata, H. (2005). Methane recovery from methane hydrate using pressurized CO_2. *Fluid Phase Equilibria, 228-229*, 553-559.

Oye, V., Aker, E., Daley, T.M., Kühn, D., Bohloli, B., Korneev, V. (2013). Microseismic monitoring and interpretation of injection data from the in Salah CO_2 storage site (Krechba), Algeria. *Energy Procedia, 37*, 4191-4198.

Palmström, A., Singh, R. (2001). The deformation modulus of rock masses—comparisons between in situ tests and indirect estimates. *Tunnelling and Underground Space Technology, 16*(2), 115-131.

Pan, F., McPherson, B.J., Dai, Z., Jia, W., Lee, S.Y., Ampomah, W., et al. (2016). Uncertainty analysis of carbon sequestration in an active CO_2-EOR field. *International Journal of Greenhouse Gas Control, 51*, 18-28.

Pan, P., Wu, Z., Feng, X., Yan, F. (2016). Geomechanical modeling of CO_2 geological storage: A review. *Journal of Rock Mechanics and Geotechnical Engineering, 8*(6), 936-947.

Pan, P.Z., Rutqvist, J., Feng, X.T., Yan, F. (2014). An approach for modeling rock discontinuous mechanical behavior under multiphase fluid flow conditions. *Rock Mechanics and Rock Engineering, 47*(2), 589-603.

Park, A.H.A., Fan, L.S. (2004). CO_2 mineral sequestration: physically activated dissolution of serpentine and pH swing process. *Chemical Engineering Science, 59*(22-23), 5241-5247.

Park, Y.C., Shinn, Y.J., Lee, H., Choi, B.Y. (2017). Assessing CO_2 storage capacity of a steeply dipping, fault bounded aquifer and effect of impurity in CO_2 stream. *Energy Procedia, 114*, 4735-4740.

Patel, M.J., May, E.F., Johns, M.L. (2017). Inclusion of connate water in enhanced gas recovery reservoir simulations. *Energy, 141*, 757-769.

Pawar, R.J., Bromhal, G.S., Carey, J.W., Foxall, W., Korre, A., Ringrose, P.S., et al. (2015). Recent advances in risk assessment and risk management of geologic CO_2 storage. *International Journal of Greenhouse Gas Control, 40*, 292-311.

Pawar, R.J., Bromhal, G.S., Chu, S., Dilmore, R.M., Oldenburg, C.M., Stauffer, P.H., et al. (2016). The National Risk Assessment Partnership's integrated assessment model for carbon storage: A tool to support decision making amidst uncertainty. *International Journal of Greenhouse Gas Control, 52*, 175-189.

Pires, J.C.M., Martins, F.G., Alvim-Ferraz, M.C.M., Simões, M. (2011). Recent developments on carbon capture and storage: An overview. *Chemical Engineering Research and Design, 89*(9), 1446-1460.

Pizzocolo, F., Peters, E., Loeve, D., Hewson, C.W., Wasch, L., Brunner, L.J. (2017). Feasibility of novel techniques to mitigate or remedy CO_2 leakage. In *SPE Europec featured at 79th EAGE Conference and Exhibition*, Paris, France, June 12-15.

Pooladi-Darvish, M., Hong, H., Theys, S.O.P., Stocker, R., Bachu, S., Dashtgard, S. (2008). CO_2 injection for enhanced gas recovery and geological storage of CO_2 in the long Coulee Glauconite F pool, Alberta. In *SPE Annual Technical Conference and Exhibition*, Denver, Colorado, UAS, September 21-24.

Preston, C., Monea, M., Jazrawi, W., Brown, K., Whittaker, S., White, D., et al. (2005). IEA GHG Weyburn CO_2 monitoring and storage project. *Fuel Processing Technology, 86*(14-15), 1547-1568.

Procesi, M., Cantucci, B., Buttinelli, M., Armezzani, G., Quattrocchi, F., Boschi, E. (2013). Strategic use of the underground in an energy mix plan: Synergies among CO_2, CH_4 geological storage and geothermal energy. Latium Region case study (Central Italy). *Applied Energy, 110*, 104-131.

Pruess, K., Battistelli, A. (2002). TMVOC, a numerical simulator for three-phase non-isothermal flows of multicomponent hydrocarbon mixtures in saturated-unsaturated heterogeneous media. Berkeley, California, USA.

Pruess, K., García, J. (2002). Multiphase flow dynamics during CO_2 disposal into saline aquifers. *Environmental Geology, 42*(2-3), 282-295.

Pruess, K., Oldenburg, C.M., Moridis, G.J. (1999). TOUGH2 user's guide version 2 (No. LBNL-43134). Berkeley, California, USA.

Rafiee, M.M., Ramazanian, M. (2011). Simulation study of enhanced gas recovery process using a compositional and a black oil simulator. In *SPE Enhanced Oil Recovery Conference*, Kuala Lumpur, Malaysia, July 19-21.

Rahman, F.A., Aziz, M.M.A., Saidur, R., Bakar, W.A.W.A., Hainin, M.R., Putrajaya, R., et al. (2017). Pollution to solution: Capture and sequestration of carbon dioxide (CO_2) and its utilization as a renewable energy source for a sustainable future. *Renewable and Sustainable Energy Reviews, 71*, 112-126.

Raju, M., Banuti, D.T., Ma, P.C., Ihme, M. (2017). Widom lines in binary mixtures of supercritical fluids. *Sci Rep, 7*(1), 3027.

Randolph, J.B., Saar, M.O. (2011). Impact of reservoir permeability on the choice of subsurface geothermal heat exchange fluid: CO_2 versus water and native brine. Paper presented at the In *the Geothermal Resources Council 35th Annual Meeting*, San Diego, California, USA, October 23-26.

Randolph, J.B., Saar, M.O. (2011). Combining geothermal energy capture with geologic carbon dioxide sequestration. *Geophysical Research Letters, 38*(10).

Raza, A., Gholami, R., Rezaee, R., Bing, C.H., Nagarajan, R., Hamid, M.A. (2018). CO_2 storage in depleted gas reservoirs: A study on the effect of residual gas saturation. *Petroleum, 4*(1), 95-107.

Raza, A., Gholami, R., Rezaee, R., Rasouli, V., Bhatti, A.A., Bing, C.H. (2018). Suitability of depleted gas reservoirs for geological CO_2 storage: A simulation study. *Greenhouse Gases: Science and Technology, 8*(5), 876-897.

Raza, A., Rezaee, R., Gholami, R., Bing, C.H., Nagarajan, R., Hamid, M.A. (2016). A screening criterion for selection of suitable CO_2 storage sites. *Journal Of Natural Gas Science And Engineering, 28*, 317-327.

Reagan, M. (2012). WebGasEOS v.2.01. Retrieved from http://esdtools.lbl.gov/gaseos/gaseos.html. http://esdtools.lbl.gov/gaseos/gaseos.html, accessed on September 19, 2019.

Reddy, K.J., John, S., Weber, H., Argyle, M.D., Bhattacharyya, P., Taylor, D.T., et al. (2011). Simultaneous capture and mineralization of coal combustion flue gas carbon dioxide (CO_2). *Energy Procedia, 4*, 1574-1583.

Regan, M.L.M. (2010). A numerical investigation into the potential to enhance natural gas recovery in water-drive gas reservoirs through the injection of CO_2. Doctoral dissertation, The University of Adelaide.

Ren, B. (2018). Local capillary trapping in carbon sequestration: Parametric study and implications for leakage assessment. *International Journal of Greenhouse Gas Control, 78*, 135-147.

Riaz, A., Cinar, Y. (2014). Carbon dioxide sequestration in saline formations: Part I—Review of the modeling of solubility trapping. *Journal of Petroleum Science and Engineering, 124*, 367-380.

Rickman, R., Mullen, M.J., Petre, J.E., Grieser, W.V., Kundert, D. (2008). A practical use of shale petrophysics for stimulation design optimization: All shale plays are not clones of the Barnett Shale. In *SPE Annual Technical Conference and Exhibition*, Denver, Colorado, USA, September 21-24.

Rinaldi, A.P., Rutqvist, J. (2013). Modeling of deep fracture zone opening and transient ground surface uplift at KB-502 CO_2 injection well, In Salah, Algeria. *International Journal of Greenhouse Gas Control, 12*, 155-167.

Rinaldi, A.P., Rutqvist, J. (2017). Modeling ground surface uplift during CO_2 sequestration: The case of in Salah, Algeria. *Energy Procedia, 114*, 3247-3256.

Rinaldi, A.P., Rutqvist, J., Finsterle, S., Liu, H.H. (2017). Inverse modeling of ground surface uplift and pressure with iTOUGH-PEST and TOUGH-FLAC: The case of CO_2 injection at In Salah, Algeria. *Computers & Geosciences, 108*, 98-109.

Ringrose, P.S., Mathieson, A.S., Wright, I.W., Selama, F., Hansen, O., Bissell, R., et al. (2013). The In Salah CO_2 storage project: Lessons learned and knowledge transfer. *Energy Procedia, 37*, 6226-6236.

Rognmo, A.U., Heldal, S., Fernø, M.A. (2018). Silica nanoparticles to stabilize CO_2-foam for improved CO_2 utilization: Enhanced CO_2 storage and oil recovery from mature oil reservoirs. *Fuel, 216*, 621-626.

Rutqvist, J. (2011). Status of the TOUGH-FLAC simulator and recent applications related to coupled fluid flow and crustal deformations. *Computers & Geosciences, 37*(6), 739-750.

Rutqvist, J., Tsang, C.F. (2005). Coupled hydromechanical effects of CO_2 injection. *Developments in Water Science, 52*, 649-679.

Rutqvist, J., Tsang, C.F. (2002). A study of caprock hydromechanical changes associated with CO_2-injection into a brine formation. *Environmental Geology, 42*(2-3), 296-305.

Rutqvist, J., Tsang, C.F. (2003). TOUGH-FLAC: a numerical simulator for analysis of coupled thermal-hydrologic-mechanical processes in fractured and porous geological media under multi-phase flow conditions. In Proceeding, TOUGH Symposium 2003, Berkeley, California, May 12-14.

Rutqvist, J., Vasco, D.W., Myer, L. (2010). Coupled reservoir-geomechanical analysis of CO_2 injection and ground deformations at In Salah, Algeria. *International Journal of Greenhouse Gas Control, 4*(2), 225-230.

Rybacki, E., Reinicke, A., Meier, T., Makasi, M., Dresen, G. (2015). What controls the mechanical properties of shale rocks? – Part I: Strength and Young's modulus. *Journal of Petroleum Science and Engineering, 135*, 702-722.

Sahin, S., Kalfa, U., Celebioglu, D. (2007). Bati Raman field immiscible CO_2 application: Status quo and future plans. In *SPE Latin American and Caribbean Petroleum Engineering Conference*, Buenos Aires, Argentina, April 15-18.

Sanders, A.W., Jones, R.M., Linroth, M., Nguyen, Q.P. (2012). Implementation of a CO_2 foam pilot study in the SACROC field: performance evaluation. In *SPE Annual Technical Conference and Exhibition*, San Antonio, Texas, USA, October 8-10.

Sanna, A., Dri, M., Maroto-Valer, M. (2013). Carbon dioxide capture and storage by pH swing aqueous mineralisation using a mixture of ammonium salts and antigorite source. *Fuel, 114*, 153-161.

Sanna, A., Uibu, M., Caramanna, G., Kuusik, R., Maroto-Valer, M.M. (2014). A review of mineral carbonation technologies to sequester CO_2. *Chemical Society Reviews, 43*(23), 8049-8080.

Sarkarfarshi, M., Malekzadeh, F.A., Gracie, R., Dusseault, M.B. (2014). Parametric sensitivity analysis for CO_2 geosequestration. *International Journal of Greenhouse Gas Control, 23*, 61-71.

Sayegh, S.G., Krause, F. F., Fosti, J. E. (1987). Miscible Displacement of Crude Oil by CO2/SO2 Mixtures. *SPE Reservoir Engineering, 2*(02), 199-208.

Schaef, H.T., McGrail, B.P., Owen, A.T. (2010). Carbonate mineralization of volcanic province basalts. *International Journal of Greenhouse Gas Control, 4*(2), 249-261.

Schepers, K.C., Nuttall, B.C., Oudinot, A.Y., Gonzalez, R.J. (2009). Reservoir modeling and simulation of the Devonian gas shale of eastern Kentucky for enhanced gas recovery and CO_2 storage. In *SPE International Conference on CO_2 Capture, Storage, and Utilization*, San Diego, California, USA, November 10-11.

Schilling, F., Borm, G., Würdemann, H., Möller, F., Kühn, M. (2009). Status report on the first European on-shore CO_2 storage site at Ketzin (Germany). *Energy Procedia, 1*(1), 2029-2035.

Schrag, D.P. (2009). Storage of carbon dioxide in offshore sediments. *Science, 325*(5948), 1658-1659.

Scripps CO_2 Program. (2019). CO_2 concentration at Mauna Loa Observatory, Hawaii. Retrieved from http://scrippsco2.ucsd.edu, accessed on June 9, 2019.

Seo, J.G. (2004). Experimental and simulation studies of sequestration of supercritical carbon dioxide in depleted gas reservoirs. Doctoral Dissertation, Texas A&M University,

Seo, J.G., Mamora, D.D. (2005). Experimental and simulation studies of sequestration of supercritical carbon dioxide in depleted gas reservoirs. *Journal of Energy Resources Technology, 127*(1), 1-6.

Seo, S., Mastiani, M., Hafez, M., Kunkel, G., Asfour, C.G., Garcia-Ocampo, K.I., et al. (2019). Injection of in-situ generated CO_2 microbubbles into deep saline aquifers for enhanced carbon sequestration. *International Journal of Greenhouse Gas Control, 83*, 256-264.

Sharma, S., Cook, P., Berly, T., Lees, M. (2009). The CO2CRC Otway Project: Overcoming challenges from planning to execution of Australia's first CCS project. *Energy Procedia, 1*(1), 1965-1972.

Shatarah, I.S., Olbrycht, R. (2017). Distributed temperature sensing in optical fibers based on Raman scattering: theory and applications. *Measurement Automation Monitoring, 63*, 41-44.

Shen, C.H., Hsieh, B.Z., Tseng, C.C., Chen, T.L. (2014). Case study of CO_2-IGR and storage in a nearly depleted gas-condensate reservoir in Taiwan. *Energy Procedia, 63*, 7740-7749.

Shi, J.Q., Smith, J., Durucan, S., Korre, A. (2013). A coupled reservoir simulation-geomechanical modelling study of the CO_2 injection-induced ground surface uplift observed at Krechba, in Salah. *Energy Procedia, 37*, 3719-3726.

Shirangi, M.G., Durlofsky, L.J. (2016). A general method to select representative models for decision making and optimization under uncertainty. *Computers & Geosciences, 96*, 109-123.

Shukla, R., Ranjith, P., Haque, A., Choi, X. (2010). A review of studies on CO_2 sequestration and caprock integrity. *Fuel, 89*(10), 2651-2664.

Hermanrud, C., Eiken, O., Hansen, O.R., Nordgård Bolås, H.M., Simmenes, T.H., Teige, G.M.G., et al. (2013). Importance of pressure management in CO_2 storage. In *Offshore Techonolgoy Conference*, Houston, Texas, USA, May 6-9.

Singh, H. (2019). Machine learning for surveillance of fluid leakage from reservoir using only injection rates and bottomhole pressures. *Journal of Natural Gas Science and Engineering, 69*, 102933.

Singh, P., Haines, M. (2014). A review of existing carbon capture and storage cluster projects and future opportunities. *Energy Procedia, 63*, 7247-7260.

Soave, G. (1972). Equilibrium constants from a modified Redlich-Kwong equation of state. *Chemical Engineering Science, 27*(6), 1197-1203.

Song, J., Zhang, D. (2013). Comprehensive review of caprock-sealing mechanisms for geologic carbon sequestration. *Environmental Science & Technology, 47*(1), 9-22.

Speight, J.G. (2007). Natural gas: a basic handbook. Gulf Publishing Company, Houston, Texas, USA.

Stein, M.H., Ghotekar, A.L., Avasthi, S.M. (2010). CO_2 sequestration in a depleted gas field: A material balance study. In *SPE EUROPEC/EAGE Annual Conference and Exhibition*, Barcelona, Spain, June 14-17.

Stork, A.L., Verdon, J.P., Kendall, J.M. (2015). The microseismic response at the In Salah Carbon Capture and Storage (CCS) site. *International Journal of Greenhouse Gas Control, 32*, 159-171.

Sun, A.Y., Jeong, H., González-Nicolás, A., Templeton, T.C. (2018). Metamodeling-based approach for risk assessment and cost estimation: Application to geological carbon sequestration planning. *Computers & Geosciences, 113*, 70-80.

Sundal, A., Hellevang, H., Miri, R., Dypvik, H., Nystuen, J.P., Aagaard, P. (2014). Variations in mineralization potential for CO_2 related to sedimentary facies and burial depth–a comparative study from the North Sea. *Energy Procedia, 63*, 5063-5070.

Székely, G.J., Rizzo, M.L., Bakirov, N.K. (2007). Measuring and testing dependence by correlation of distances. *The Annals of Statistics, 35*(6), 2769-2794.

Tan, Y., Nookuea, W., Li, H., Thorin, E., Yan, J. (2016). Property impacts on Carbon Capture and Storage (CCS) processes: A review. *Energy Conversion and Management, 118*, 204-222.

Tanaka, K., Vilcáez, J., Sato, K. (2013). Improvement of CO_2 geological storage efficiency by injection and production well design. *Energy Procedia, 37*, 4591-4597.

Tang, Y., Yang, R., Bian, X. (2014). A review of CO_2 sequestration projects and application in China. *Scientific World Journal, 2014*, 381854.

Tapia, J.F.D., Lee, J.Y., Ooi, R.E.H., Foo, D.C.Y., Tan, R.R. (2014). CO_2 allocation for scheduling enhanced oil recovery (EOR) operations with geological sequestration using discrete-time optimization. *Energy Procedia, 61*, 595-598.

Teatini, P., Castelletto, N., Ferronato, M., Gambolati, G., Janna, C., Cairo, E., et al. (2011). Geomechanical response to seasonal gas storage in depleted reservoirs: A case study in the Po River basin, Italy. *Journal of Geophysical Research: Earth Surface, 116*, F02002.

Teir, S., Eloneva, S., Fogelholm, C.J., Zevenhoven, R. (2009). Fixation of carbon dioxide by producing hydromagnesite from serpentinite. *Applied Energy, 86*(2), 214-218.

Tek, M.R. (1989). Underground storage of natural gas: theory and practice (Vol. 171). Springer Science & Business Media, Berlin, Germany.

Teng, Y., Zhang, D. (2018). Long-term viability of carbon sequestration in deep-sea sediments. *Science advances, 4*(7), eaao6588.

Tseng, C.C, Chen, D.L., Hu, S.T., Lin, Z.S. (2011). A simulation study on CO_2 storage pilot test for the YHS Field (in Chinese). *Mining & Metallurgy, 56*(1), 23-40.

Turta, A.T., Sim, S.S., Singhal, A.K., Hawkins, B.F. (2007). Basic investigations on enhanced gas recovery by gas-gas displacement. In *Petroleum Society's 8th Canadian International Petroleum Conference (58th Annual Technnical Meeting)*, Calgary, Alberta, Canada, June 12-14.

Twerda, A., Belfroid, S., Neele, F. (2018). CO_2 injection in low pressure depleted reservoirs. In the *Fifth CO_2 Geological Storage Workshop*, Utrecht, The Netherlands, November 21-23.

Underhill, D.H. Analysis of uranium supply to 2050. Retrieved from https://inis.iaea.org/search/search.aspx?orig_q=RN:31054412, accessed on October 16, 2019.

Van Genuchten, M.T. (1980). A closed-form equation for predicting the hydraulic conductivity of unsaturated soils. *Soil Science Society of America Journal, 44*(5), 892-898.

Vandeweijer, V., Hofstee, C., Graven, H. (2018). 13 years of safe CO_2 injection at K12-B. In the *Fifth CO_2 Geological Storage Workshop*, Utrecht, Netherlands, November 21-23.

Vandeweijer, V., van der Meer, B., Hofstee, C., Mulders, F., D'Hoore, D., Graven, H. (2011). Monitoring the CO_2 injection site: K12-B. *Energy Procedia, 4*, 5471-5478.

Vapnik, V. (2013). The nature of statistical learning theory. Springer science & business media.

Vargaftik, N.B. (1975). Tables on the thermophysical properties of liquids and gases in normal and dissociated states (2nd Ed.). John Wiley & Sons, New York, UAS.

Verduyn, M., Geerlings, H., van Mossel, G., Vijayakumari, S. (2011). Review of the various CO_2 mineralization product forms. *Energy Procedia, 4*, 2885-2892.

Vu, H.P., Black, J.R., Haese, R.R. (2018). The geochemical effects of O_2 and SO_2 as CO_2 impurities on fluid-rock reactions in a CO_2 storage reservoir. *International Journal of Greenhouse Gas Control, 68*, 86-98.

Walker, W.R., Sabey, J.D., Hampton, D.R. (1981). Studies of heat transfer and water migration in soils. Final report, Departement of Agricultural and Chemical Engineering, Colorado State University, Fort Collins.

Wang, H., Wang, J.G. (2018). A fully coupled mathematical model with geochemical reaction for caprock sealing efficiency in CO_2 geosequestration. In *CHINA ROCK 2018-The 15th China Rock Mechanics and Engineering Academic Annual Meeting*, Beijing, China, Semptember 19-22.

Wang, J.G., Liu, J., Liu, J., Chen, Z. (2013). Impact of rock microstructures on the supercritical CO_2 enhanced gas recovery. In *CPS/SPE International Oil & Gas Conference and Exhibition in China*, Beijing, China, June 8-10.

Wang, J., Ryan, D., Anthony, E.J., Wigston, A., Basava-Reddi, L., Wildgust, N. (2012). The effect of impurities in oxyfuel flue gas on CO_2 storage capacity. *International Journal of Greenhouse Gas Control, 11*, 158-162.

Wang, J., Ryan, D., Anthony, E.J., Wildgust, N., Aiken, T. (2011). Effects of impurities on CO_2 transport, injection and storage. *Energy Procedia, 4*, 3071-3078.

Wang, X., Economides, M. (2009). Advanced natural gas engineering. Gulf Publishing Company, Houston, Texas, USA.

Wang, X., Economides, M.J. (2012). Purposefully built underground natural gas storage. *Journal of Natural Gas Science and Engineering, 9*, 130-137.

Wang, X., Maroto-Valer, M.M. (2013). Optimization of carbon dioxide capture and storage with mineralisation using recyclable ammonium salts. *Energy, 51*, 431-438.

Wasch, L.J., Wollenweber, J., Tambach, T.J. (2013). Intentional salt clogging: a novel concept for long-term CO_2 sealing. *Greenhouse Gases: Science and Technology, 3*(6), 491-502.

Wee, J.H. (2013). A review on carbon dioxide capture and storage technology using coal fly ash. *Applied Energy, 106*, 143-151.

Wei, N., Li, X., Fang, Z., Bai, B., Li, Q., Liu, S., et al. (2015). Regional resource distribution of onshore carbon geological utilization in China. *Journal of CO_2 Utilization, 11*, 20-30.

Wei, N., Li, X., Wang, Y., Wang, Y., Kong, W. (2013). Numerical study on the field-scale aquifer storage of CO_2 containing N_2. *Energy Procedia, 37*, 3952-3959.

Wei, N., Li, X., Wang, Y., Zhu, Q., Liu, S., Liu, N., et al. (2015). Geochemical impact of aquifer storage for impure CO_2 containing O_2 and N_2: Tongliao field experiment. *Applied Energy, 145*, 198-210.

Williams, G., Chadwick, A. (2018). Chimneys and channels: History matching the growing CO_2 plume at the Sleipner storage site. In the *Fifth CO_2 Geological Storage Workshop*, Utrecht, The Netherlands, November 21-23.

Wolf, J.L., Fischer, S., Rütters, H., Rebscher, D. (2017). Reactive transport simulations of impure CO_2 injection into saline aquifers using different modelling approaches provided by TOUGHREACT V3.0-OMP. *Procedia Earth and Planetary Science, 17*, 480-483.

Wolf, J.L., Niemi, A., Bensabat, J., Rebscher, D. (2016). Benefits and restrictions of 2D reactive transport simulations of CO_2 and SO_2 co-injection into a saline aquifer using TOUGHREACT V3.0-OMP. *International Journal of Greenhouse Gas Control, 54*, 610-626.

Worthen, A.J., Bagaria, H.G., Chen, Y., Bryant, S.L., Huh, C., Johnston, K.P. (2013). Nanoparticle-stabilized carbon dioxide-in-water foams with fine texture. *Journal of Colloid and Interface Science, 391*, 142-151.

Worthen, A.J., Bryant, S.L., Huh, C., Johnston, K.P. (2013). Carbon dioxide-in-water foams stabilized with nanoparticles and surfactant acting in synergy. *AIChE Journal, 59*(9), 3490-3501.

Worthen, A.J., Parikh, P.S., Chen, Y., Bryant, S.L., Huh, C., Johnston, K.P. (2014). Carbon dioxide-in-water foams stabilized with a mixture of nanoparticles and surfactant for CO_2 storage and utilization applications. *Energy Procedia, 63*, 7929-7938.

Wriedt, J., Deo, M., Han, W.S., Lepinski, J. (2014). A methodology for quantifying risk and likelihood of failure for carbon dioxide injection into deep saline reservoirs. *International Journal of Greenhouse Gas Control, 20*, 196-211.

Wu, B., Li, X., Mutailipu, M., Yang, M., Wang, D., Jiang, L., et al. (2017). Saturation comparison of CO_2 containing N_2 at different injection velocities and conditions by X-ray CT scanning. *Energy Procedia, 142*, 3350-3355.

Xu, C.G., Cai, J., Yu, Y.S., Chen, Z.Y., Li, X.S. (2018). Research on micro-mechanism and efficiency of CH_4 exploitation via CH_4-CO_2 replacement from natural gas hydrates. *Fuel, 216*, 255-265.

Xu, Z., Fang, Y., Scheibe, T.D., Bonneville, A. (2012). A fluid pressure and deformation analysis for geological sequestration of carbon dioxide. *Computers & Geosciences, 46*, 31-37.

Yang, F., Zhao, G.B., Adidharma, H., Towler, B., Radosz, M. (2007). Effect of oxygen on minimum miscibility pressure in carbon dioxide flooding. *Industrial & Engineering Chemistry Research, 46*(4), 1396-1401.

Yang, G., Li, Y., Atrens, A., Liu, D., Wang, Y., Jia, L., et al. (2017). Reactive transport modeling of long-term CO_2 sequestration mechanisms at the Shenhua CCS demonstration project, China. *Journal of Earth Science, 28*(3), 457-472.

Yang, W., Peng, B., Liu, Q., Wang, S., Dong, Y., Lai, Y. (2017). Evaluation of CO_2 enhanced oil recovery and CO_2 storage potential in oil reservoirs of Bohai Bay Basin, China. *International Journal of Greenhouse Gas Control, 65*, 86-98.

Yaws, C.L. (1999). Chemical properties handbook: physical, thermodynamical, environmental, transport, safety and health related properties for organic and inorganic chemicals. McGaw-Hill, New York, USA.

You, J., Ampomah, W., Kutsienyo, E.J., Sun, Q., Balch, R.S., Aggrey, W.N., et al. (2019). Assessment of enhanced oil recovery and CO_2 storage capacity using machine learning and optimization framework. In *SPE Europec featured at 81st EAGE Conference and Exhibition*, London, UK, June 3-6.

Yu, H., Zhou, G., Fan, W., Ye, J. (2007). Predicted CO_2 enhanced coalbed methane recovery and CO_2 sequestration in China. *International Journal of Coal Geology, 71*(2-3), 345-357.

Yuan, Q., Sun, C.Y., Yang, X., Ma, P.C., Ma, Z.W., Liu, B., et al. (2012). Recovery of methane from hydrate reservoir with gaseous carbon dioxide using a three-dimensional middle-size reactor. *Energy, 40*(1), 47-58.

Zahid, U., Lim, Y., Jung, J., Han, C. (2011). CO_2 geological storage: A review on present and future prospects. *Korean Journal of Chemical Engineering, 28*(3), 674-685.

Zangeneh, H., Safarzadeh, M.A. (2017). Enhanced gas recovery with carbon dioxide sequestration in a water-drive gas condensate reservoir: a case study in a real gas field. *Journal of Petroleum Science and Technology, 7*(2), 3-11.

Zangeneh, H., Jamshidi, S., Soltanieh, M. (2013). Coupled optimization of enhanced gas recovery and carbon dioxide sequestration in natural gas reservoirs: Case study in a real gas field in the south of Iran. *International Journal of Greenhouse Gas Control, 17*, 515-522.

Zhang, D., Ranjith, P.G., Perera, M.S.A. (2016). The brittleness indices used in rock mechanics and their application in shale hydraulic fracturing: A review. *Journal of Petroleum Science and Engineering, 143*, 158-170.

Zhang, G., Li, B., Zheng, D., Ding, G., Wei, H., Qian, P., et al. (2017). Challenges to and proposals for underground gas storage (UGS) business in China. *Natural Gas Industry B, 4*(3), 231-237.

Zhang, K., Wu, Y.S., Pruess, K. (2008). User's guide for TOUGH2-MP-a massively parallel version of the TOUGH2 code (No. LBNL-315E). Berkeley, California, USA.

Zhang, L., Li, X., Zhang, Y., Cui, G., Tan, C., Ren, S. (2017). CO_2 injection for geothermal development associated with EGR and geological storage in depleted high-temperature gas reservoirs. *Energy, 123*, 139-148.

Zhang, L., Niu, B., Ren, S., Zhang, Y., Yi, P., Mi, H., et al. (2010). Assessment of CO_2 storage in DF1-1 gas field South China Sea for a CCS demonstration. *Journal of Canadian Petroleum Technology, 49*(08), 9-14.

Zhang, L., Yang, L., Wang, J., Zhao, J., Dong, H., Yang, M., et al. (2017). Enhanced CH_4 recovery and CO_2 storage via thermal stimulation in the CH_4/CO_2 replacement of methane hydrate. *Chemical Engineering Journal, 308*, 40-49.

Zhang, M., Bachu, S. (2011). Review of integrity of existing wells in relation to CO_2 geological storage: What do we know? *International Journal of Greenhouse Gas Control, 5*(4), 826-840.

Zhang, P.Y., Huang, S., Sayegh, S., Zhou, X.L. (2004). Effect of CO_2 impurities on gas-injection EOR processes. In *SPE/DOE Symposium on Improved Oil Recovery*, Tulsa, Oklahoma, USA.

Zhang, X., Li, Q., Zheng, L., Li, X., Xu, L. (2020). Numerical simulation and feasibility assessment of acid gas injection in a carbonate formation of the Tarim Basin, China. *Oil & Gas Science and Technology–Rev. IFP Energies nouvelles*, 75, 28.

Zhang, Z., Agarwal, R.K. (2012). Numerical simulation and optimization of CO_2 sequestration in saline aquifers for vertical and horizontal well injection. *Computational Geosciences, 16*(4), 891-899.

Zhang, Z., Agarwal, R. (2013). Numerical simulation and optimization of CO_2 sequestration in saline aquifers. *Computers & Fluids, 80*, 79-87.

Zhang, Z., Huisingh, D. (2017). Carbon dioxide storage schemes: Technology, assessment and deployment. *Journal of Cleaner Production, 142*, 1055-1064.

Zhao, X., Liao, X., Wang, W., Chen, C., Rui, Z., Wang, H. (2014). The CO_2 storage capacity evaluation: Methodology and determination of key factors. *Journal of the Energy Institute, 87*(4), 297-305.

Ziabakhsh-Ganji, Z., Kooi, H. (2014). Sensitivity of Joule–Thomson cooling to impure CO_2 injection in depleted gas reservoirs. *Applied Energy, 113*, 434-451.

Zuloaga, P., Yu, W., Miao, J., Sepehrnoori, K. (2017). Performance evaluation of CO_2 Huff-n-Puff and continuous CO_2 injection in tight oil reservoirs. *Energy, 134*, 181-192.

www.ingramcontent.com/pod-product-compliance
Ingram Content Group UK Ltd.
Pitfield, Milton Keynes, MK11 3LW, UK
UKHW022001190726
13853UKWH00004B/1663

9 783736 973862